Water Scarcity, Livelihoods and Food Security

This volume reviews the evolution of ten years' learning and discovery about water scarcity, livelihoods and food security within the CGIAR Challenge Program on Water and Food. It draws on the experiences of over 100 projects conducted in ten river basins in the developing world.

The book describes how the program's design evolved from an emphasis on water scarcity, water productivity and water access to an emphasis on using water innovations to improve livelihoods and address development challenges in specific river basins. It shows how the research was used to foster change in stakeholder behavior, linking it to improved knowledge, attitudes and skills, which were fostered by stakeholder participation, innovation, dialogue and negotiation.

The authors describe development challenges, their drivers and their political context; how to address them through technical, institutional and policy innovations; and the consequences of change at different scales and time frames on equity, resilience and ecosystem services. Overall, the work represents a major synthesis and landmark publication for all concerned with water resource management and sustainable development.

Larry W. Harrington was Research Director, Challenge Program on Water and Food (CPWF) of CGIAR, based at the International Water Management Institute (IWMI), Colombo, Sri Lanka, now at Ithaca, NY, USA.

Myles J. Fisher is an Emeritus Scientist, Centro Internacional de Agricultura Tropical (CIAT), Cali, Colombia.

'From a water scarcity and productivity programme to one of development challenges, this book presents the commendable work of the CPWF in very diverse basins all over the world. The processes implemented built on participation, innovation, dialogue and negotiation among the locals, ultimate users of the resource, will ensure the positive legacy of the programme. Since lessons learnt will stay with the locals, and will not leave with the donors, the initial expectations of research for development may even be surpassed. Certainly a lesson for many donors.'

Cecilia Tortajada, President, Third World Centre for Water Management, Mexico.

Earthscan Studies in Water Resource Management

For more information and to view forthcoming titles in this series, please visit the Routledge website: http://www.routledge.com/books/series/ECWRM/

Water Scarcity, Livelihoods and Food Security

Research and innovation
for development

Edited by Larry W. Harrington and Myles J. Fisher

First published 2014
by Routledge
2 Park Square, Milton Park, Abingdon, Oxon OX14 4RN

and by Routledge
711 Third Avenue, New York, NY 10017

Routledge is an imprint of the Taylor & Francis Group, an informa business

© 2014 International Water Management Institute

British Library Cataloguing-in-Publication Data
A catalogue record for this book is available from the British Library

Library of Congress Cataloging-in-Publication Data
Water scarcity, livelihoods and food security : research and innovation for development / edited by Larry W. Harrington and Myles J. Fisher.
     pages cm. – (Earthscan studies in water resource management)
     Includes bibliographical references and index.
  1. Water resources development–International cooperation. 2. Water-supply–International cooperation. 3. Food security–International cooperation. 4. Challenge Program on Water and Food. 5. Agriculture–Research–International cooperation. I. Harrington, Larry W. II. Fisher, Myles. J.
     HD1691.W363 2014
     333.91–dc23   2014007179

ISBN: 978-0-415-72846-1(hbk)
ISBN: 978-0-415-72847-8 (pbk)
ISBN: 978-1-315-85166-2 (ebk)

Typeset in Bembo
by Keystroke, Station Road, Codsall, Wolverhampton

Contents

viii *Contents*

Foreword

The CGIAR Challenge Program on Water and Food (CPWF) was designed as a 15-year program (2004–2018) that addressed interrelated issues of water scarcity, water productivity, livelihoods, food security, poverty and the environment. It was conceived as a response by the CGIAR to a perceived global crisis of water scarcity and the urgent need to use increasingly scarce water resources more efficiently. With the passage of time, the CPWF broadened its agenda to focus on a range of water-related development challenges in river basins. The CPWF came to see that water provides a useful, even essential, entry point for addressing many development challenges. These included challenges related to the sustainable intensification of agricultural systems and preserving ecosystem services where these also see positive changes in rural people's poverty. Through the latest CGIAR reform, the duration of the Program was shortened to ten years.

The CPWF was designed, as were three other CGIAR Challenge Programs, to explore new ways of doing research with partners for development purposes. In its more than 50-year history, the CGIAR has had long experience with cross-Center initiatives (Ecoregional Programs, System-wide Programs, Challenge Programs, and more recently CGIAR Research Programs or CRPs). These have helped the whole system progress despite recurrent financial insecurity and uncertainties arising from multiple rounds of institutional change. The CPWF was not immune to this type of turbulence.

This is a good time for us to express our admiration and sincere thanks to the project teams, and to their institutions and partners, for their perseverance, their inspiration and their cooperation. In a sustained and enthusiastic effort, they have demonstrated adaptability and a willingness to change, producing the scientific outputs and the development outcomes presented in this book.

In research-for-development, scientific results are most useful when they are credible and relevant, and when they inform engagement with policy- and decision-makers. An important measure of our success lies therefore in the role those results will play in future decisions.

In preparing this book, we aimed to provide in one place a top-level summary of what has been learned through the CPWF experience covering Phase 1, the Basin Focal Projects and Phase 2. The book tells the evolution of ten years' learning and discovery about water scarcity, livelihoods, and food security within the CPWF. It draws on the experience of 120 projects conducted in ten river basins in the developing world. It describes how the program's design started from an emphasis on water scarcity, water productivity and water access. That design evolved to an emphasis on using water innovations to address development challenges in specific river basins. It tells how CPWF used research to foster change in stakeholder behavior, linking it to improved knowledge, attitudes and skills. These were fostered by stakeholder participation, innovation, dialogue and negotiation. It describes development challenges, their drivers and their political context, and how to address them through technical, institutional and policy innovations.

The book features nine chapters. We first review the origin of CGIAR Global Challenge Programs and describe the evolution of the CPWF (Chapter 1) within its wider global context. Then we revisit the concepts that drove the original program, especially water scarcity and water productivity (Chapter 2) and show how a reconsideration of these concepts led us to introduce "research for development" (R4D) to address development challenges or "wicked problems" in river basins (Chapter 3). We present an institutional history of the CPWF, recording key events or factors that influenced the Program's way of working (Chapter 4). We next discuss research on technologies, exploring the complementarity between technical and institutional innovation, and placing this research within our theory of change (Chapter 5). We then describe the contribution of CPWF research to understand how water research management institutions work, how they influence water allocation and use, and how they can be strengthened (Chapter 6). After this, we discuss how partnerships and innovation platforms helped generate information aimed at influencing decision-making or negotiations and explore the influence of power relationships on R4D processes (Chapter 7). Finally, we pull together the threads from previous chapters to present specific instances of using R4D to get from research outputs to development outcomes in specific instances (Chapter 8) and summarize basin-level and program-level messages (Chapter 9).

This book is directed at several audiences, among them researchers interested in development and development workers interested in the contributions of research to problem solving. We also target research managers from national and international institutions, donor and development assistance agencies, NGOs, students and young scientists, the CPWF community, and the CGIAR and its Research Programs.

The book describes practical lessons from a R4D community. We consider that its successes and its promise open up many opportunities for future investment by donors. In particular we see how broad partnerships and a focus on useable products often succeeded in producing results that went beyond business-as-usual and made a real difference on the ground. We also see how younger researchers sensed the value of their work for society in general and for the poor in particular, and as a result committed themselves with energy.

As the CPWF comes to an end, it is clear that not only CPWF but all R4D programs operate in highly political environments, which can be more or less enabling, requiring long time frames, trust and adaptability. And we know from our CPWF experience, for example in the Andes region or in the Limpopo basin in Southern Africa, that policy change often needs 10–20 years to unfold. Hence much remains to be done to move from the early outcomes that CPWF projects have begun to generate towards achieving impacts on the ground. This will require continued effort from global, regional and local research and development partners, as well as intelligent choices by donors about investment in the opportunities that emerge from CPWF. We are glad to see that FANRPAN, CONDESAN and three CGIAR Research Programs (HumidTropics, AAS—Aquatic Agricultural Systems, and moreover WLE— Water Land and Ecosystems) have already embarked on this effort.

Jonathan Woolley, CPWF Director (2003–2009)
Alain Vidal, CPWF Director (2009–2014)

Abbreviations

AET	actual evapotranspiration
BDCs	Basin Development Challenges
BFPs	Basin Focal Projects
BSM	benefit-sharing mechanism
CA	*Comprehensive Assessment of Water in Agriculture* (Molden 2007, ed.)
CAC	*Conversatorio de Acción Ciudadana*
CB	Consortium Board
CDMT	Change Design and Management Team
CGIAR	Consultative Group on International Agricultural Research (up to 2010. Then restructured and renamed 'CGIAR Consortium of International Agricultural Research Centers')
CIAT	Centro Internacional de Agricultura Tropical (International Center for Tropical Agriculture)
ComMod	companion modeling
CONDESAN	*Consortio para del Desarrollo Sonstenible de la Ecoregión Andina*
CP	Challenge Program
CPWF	Challenge Program on Water and Food
CRP	CGIAR Research Program
CSC	Consortium Steering Committee
DSS	decision-support system
DWAF	Department of Water Affairs and Forestry (South Africa)
ESSs	ecosystem services
ET	evapotranspiration
ExCo	Executive Council
FANRPAN	Food, Agriculture and Natural Resources Policy Analysis Network
FAO	United Nations Food and Agriculture Organization
GCPs	Global Challenge Programs
IAD	Institutional analysis and development
IDE	International Development Enterprises
IFPRI	International Food and Policy Research Institute

INRM	integrated natural resource management
iSC	interim Science Council
IWMI	International Water Management Institute
IWRM	integrated water resource management
KAS	knowledge, attitudes and skills
KM	knowledge management
LEDs	local engineering departments
LWP	livestock water productivity
M&E	monitoring and evaluation
M-POWER	Mekong Program on Water, Environment and Resilience
MT	management team
MUS	multiple-use water services
NARES	national agricultural research and extension systems
NARS	national agricultural research systems
NDI	Nepal Department of Irrigation
OLM	outcome logic model
PAM	policy analysis matrix
PESs	payments for ecosystem services
PET	potential evapotranspiration
PIPA	participatory impact pathway analysis
QSMAS	Quesungual slash and mulch agroforestry system
R4D	research for development
SADC	Southern African Development Community
SCALES	sustaining collective action that links across economic and ecological scales
SGPs	small grants projects
SLF	sustainable livelihood framework
SLM	sustainable land management
SRPs	strategic research portfolios
TAC	Technical Advisory Committee
ToC	theory of change
TWGs	Topic Working Groups
WLE	Water, Land and Ecosystems (CGIAR Research Program)
WP	water productivity
WSSD	World Summit on Sustainable Development

1 The Challenge Program on Water and Food: A new paradigm for research in the CGIAR[1]

Myles J. Fisher,[a]* *Amanda Harding*[b]
and *Eric Kemp-Benedict*[c]

[a]Centro Internacional de Agricultura Tropical CIAT, Cali, Colombia;
[b]CGIAR Challenge Program on Water and Food CPWF, Paris, France;
[c]Stockholm Environment Institute SEI, Bangkok, Thailand; *Corresponding author, mylesjfisher@gmail.com.

The creation of Global Challenge Programs and the Challenge Program on Water and Food (CPWF)

The Consultative Group on International Agricultural Research (CGIAR) has always been about food security. It started over 40 years ago in 1971 with four Centers focused on breeding better staple food crops. In 2000, when it consisted of 16 Centers, it asked its Technical Advisory Committee (TAC) to address its future for the next decade, what it should be doing and producing; how it should be doing it and with whom. TAC produced *A Food Secure World for All: Toward a New Vision and Strategy for the CGIAR*[2] to guide it through the coming decade (Box 1.1), which was approved at International Centers' Week, 2000.[3]

The CGIAR Chair commissioned a *Change Design and Management Team* (CDMT) to make concrete proposals for how TAC's proposals might be implemented. The CDMT recommended that "[The] CGIAR should formulate

---

Box 1.1 TAC's vision and strategy

Vision: A food secure world for all.

Goal: To reduce poverty, hunger and malnutrition by sustainably increasing the productivity of resources in agriculture, forestry and fisheries.

Mission: To achieve sustainable food security and reduce poverty in developing countries through scientific research and research-related activities in the fields of agriculture, livestock, forestry, fisheries, policy and natural resources management.

and implement a few ... Global Challenge Programs (GCPs), which are focused on specific outputs and are based on an inclusive approach to priority setting ... They should be funded significantly by additional resources."

One possible GCP identified by the CDMT was "Improved water management practices for agriculture." Although this set the stage for the submission of a GCP on water in agriculture, there were already powerful movements towards increasing global recognition of the critical state of water, food production and poverty.

World Water Council activities, 1998–2000

In 1997, the World Water Council created a *long-term vision on water, life and the environment in the 21st century* (Cosgrove and Rijsberman, 1998), which detailed a comprehensive series of activities leading up to the 2nd World Water Forum and a parallel Ministerial Conference in The Hague in 2000. Amongst the activities, which were "meant to move us from where we are today to where we need to be to meet future water needs and ensure the sustainable use of water," were consultations to obtain visions of the needs for "*water for food* (including both rainfed and irrigated agriculture" (Cosgrove and Rijsberman, 1998), emphasis is from the original paper). The 2nd World Water Forum, with 5500 delegates, and the parallel Ministerial Conference, with 600 delegates, including 120 ministers, were major international events. Their recommendations influenced subsequent deliberations in the CGIAR and elsewhere.

The Challenge Program on Water and Food, justification and intent

In early 2002, the CGIAR interim Science Council (iSC, which superseded the TAC), chose the Challenge Program on Water and Food (CPWF) together with two others[4] to go forward for development as full proposals by mid-year. The full proposal of the CPWF was, "an ambitious research, extension and capacity building program that will significantly increase the productivity of water used for agriculture ... in a manner that is environmentally sustainable and socially acceptable." The intermediate objective was

> [T]o maintain the level of global diversions of water to agriculture at the level of the year 2000, while increasing food production, to achieve internationally adopted targets for decreasing malnourishment and rural poverty by the year 2015, particularly in rural and peri-urban areas in river basins with low average incomes and high physical, economic or environmental water scarcity or water stress, with a specific focus on low-income groups within these areas.

The iSC endorsed the CPWF proposal at the end of August 2002 for approval by the Executive Council (ExCo). ExCo endorsed the proposal and recommended its approval by the CGIAR on 22 September 2002. ExCo noted that

"The proposal demonstrates clear linkages with global work on water and food, demonstrates wide stakeholder inclusion, national agricultural research systems (NARS) participation is very high, and other partners are well represented."

After the iSC endorsement, the World Summit on Sustainable Development (WSSD) was held in Johannesburg 26 August–4 September 2002. The WSSD produced the Johannesburg Plan of Implementation, of which paragraph 40 states,

> Agriculture plays a crucial role in addressing the needs of a growing global population and is inextricably linked to poverty eradication, especially in developing countries. Enhancing the role of women at all levels and in all aspects of rural development, agriculture, nutrition and food security is imperative. Sustainable agriculture and rural development are essential to the implementation of an integrated approach to increasing food production and enhancing food security and food safety in an environmentally sustainable way.

Subparagraph 40(d) reads, "Promote programmes to enhance in a sustainable manner the productivity of land and the efficient use of water resources in agriculture, forestry, wetlands, artisanal fisheries and aquaculture, especially through indigenous and local community-based approaches."

Paragraph 40 provided the policy legitimacy for the research directions and themes of the CPWF: it can be seen from two sides. The global water community needed to address the issue of water management in agriculture within the context of finite water resources under increasing pressure. The agricultural sector needed to identify ways to enhance resource productivity in agriculture, including water productivity. This view supported the establishment of the CPWF as a worldwide program aimed at increasing water productivity in agriculture from the community to whole basin scales.

The focus on water productivity remained foremost in the thinking of the CPWF for several years after its inception. "The most important question in the current debate on water scarcity is not so much whether it is true or not, whether we are going to run out of water or not, whether water scarcity is fact or fiction, but whether this debate will help increase water productivity" (Rijsberman, 2004).

CPWF context within the CGIAR's new programmatic approach

As intended by the CDMT, the CPWF introduced a new model for research for the CGIAR with the emphasis on collaboration, both between Centers, and between Centers and national agricultural research and extension systems (NARES) and advanced research institutes. When appropriate, the new model used a participatory, integrated natural resource management (INRM) approach to develop and disseminate technology (Sayer and Campbell, 2003).

The CPWF emphasized team work in which all participants shared knowledge and which led to technological innovation.

The GCPs did not exist as independent fiduciary entities, so that the CPWF operated under the umbrella of the CGIAR International Water Management Institute (IWMI). This led to administrative anomalies, such as the program coordinator reporting to the Consortium Steering Committee, while IWMI management evaluated the coordinator's performance. Similarly, the program coordinator had little authority over CPWF management staff, who were employed and evaluated by the different consortium institutions involved.

Incremental funding

The CDMT foresaw that as more GCPs were created, they could together require as much as 50 percent of the CGIAR's budget. The iSC recognized early on that this was unlikely and, although not stated, would certainly meet fierce resistance from the Centers and those donors aligned to particular Centers. The iSC believed that, "The Centers expect the [Challenge Program] funding to be new and incremental . . ."[5] and proposed that the GCPs should seek new funding, which would add to the system's total budget.

The CPWF secured new funding of nearly US$70 million for 2003–2008 from a broad spectrum of donors, which gave it independence from individual donors. It also managed to compensate partly "for a drastic reduction of a major donor commitment in the programme inception phase,"[6] US$25 million to only US$5 million when the government of the Netherlands changed in May 2003.

Water and food sub-systems

The aim of the CPWF was to increase water productivity through better management of water for food production. The CPWF identified three levels of system organization. At the lowest level, the plant–field–farm system, there are three sub-systems, agroecosystems, upper catchments, and aquatic eco-systems. The second level is the river basin, where different water users interact, and where the trade-offs between and among water users are impor-tant. These determine the interactions between surface water, groundwater, and precipitation as well as the interactions between upstream and downstream users. The third level is the national and global water and food systems. The external environment was considered at all levels, including not only the water sector, but the macroeconomic factors that impact it, as well as policies and institutional issues at global and national levels.

Research themes

The three sub-systems of the lowest level plus the basin and global levels coincide with the five research themes that the CPWF identified (Box 1.2).

Box 1.2 The five original research themes of CPWF Phase 1 (lead Center)

Theme 1: Improve crop water productivity (IRRI).
Theme 2: Multiple use of upper catchments (CIAT).
Theme 3: Aquatic ecosystems and fisheries (WorldFish).
Theme 4: Integrated basin water management systems (IWMI).
Theme 5: The global and national food and water system (IFPRI).

Note: IRRI = International Rice Research Institute; CIAT = Centro Internacional de Agricultura Tropical; IWMI = International Water Management Institute; IFPRI = International Food and Policy Research Institute.

The research themes were given a geographic focus carrying out research in one or more of nine benchmark basins.

The theme to improve crop water productivity included a wide range of crops, environments, scale levels, and methodologies varying from biotechnology to geographic information systems, and remote sensing.

Multiple use of upper catchments explored ways to improve the use of water and other resources by understanding the relationships between water, livelihoods and poverty at multiple scales. The objective was to design interventions that are both sustainable and equitable.

Aquatic ecosystems and fisheries are important in the livelihoods of many of the world's poor, for example supplying 60 percent of dietary protein in Cambodia. The theme focused on assessing the economic value of aquatic ecosystem goods and services; integrating crops with aquaculture, and improving the management of fisheries in reservoirs.

The theme on basin-level water management focused on analysis of water productivity in rain-fed and irrigated farming systems. The objective was to identify basin-level interventions that enhance human and ecological well-being by increasing water productivity.

The theme on global food and water systems developed a conceptual framework to analyze food production systems at national and global scales to identify strengths and weaknesses in the use of green and blue water. It used two approaches: (a) scenario analysis, including drivers and development goals; and (b) stakeholder participatory research and institutional analysis.

Following a worldwide call, over 400 research project proposals were received of which 55 were finally approved, following a stringent evaluation process. Five Theme Leaders and nine Basin Coordinators based in different institutions acted as the management team, providing oversight to link technical quality with support for out- and up-scaling and to ensure the quality of the contracted projects in the nine benchmark basins (Box 1.3). The purpose

---

Box 1.3 Benchmark basins

South America: A group of small basins in the Andes, São Francisco.
Africa: Volta, Limpopo, Nile.
Asia: Karkheh, Indus-Ganges, Mekong, Yellow.

---

of the benchmark basins was to integrate research across themes at the basin level by working closely with stakeholders and prioritizing the research most relevant to each basin. Teams within each basin developed baselines against which progress and impacts were assessed.

Toward the end of Phase 1, the iSC criticized the lack of geographical and thematic coherence in the first round of 55 approved projects. In response, the Consortium Steering Committee created Basin Focal Projects (BFPs) to present a globally coherent picture of whole-basin systems that recognized the large differences in hydrology (and consequent livelihood systems) within and between basins. The work of the BFP teams was to show the link between poverty, agriculture and water within each benchmark basin, and to develop rigorous conceptual frameworks to enable scientists to analyze these links in other river basins at various scales of resolution. The CPWF responded to an external review commissioned by the iSC and to changes within the CGIAR by shifting the focus away from research outputs, to an emphasis on broader outcomes produced as a result of research. We discuss this evolution below.

Water, development and poverty

During the initial phase (2003–2007), CPWF research for development (R4D) was in the context of diverse, water-related problems and focused on identifying and selecting what strategies had most potential to improve food security and reduce poverty. As the CPWF gained understanding of the complex relationships between agricultural water management and poverty—and the dynamics of water, food, and poverty—it saw that the level of socio-economic development was a key driver. It also saw that the natural-resources management (NRM) approach it was using was well suited for research into many development issues (World Commission on Environment and Development, 1987). The second phase of the CPWF therefore focused on alleviating poverty and increasing farmers' and farming systems' resilience, which is often driven by external global forces at different spatial and institutional levels, such as shocks to financial markets and climate change.

Poverty and development: the broader context

Many scenarios of the future forecast conflict (The 2030 Water Resources Group, 2009; Deloitte, 2012; KPMG, 2012; McKinsey and Company, 2012) and conclude that food and environmental insecurity and poverty will be widespread, paying little attention to constructive solutions such as adaptation and innovation. In contrast, in the second phase the CPWF addressed these issues in a wider global context. It researched the drivers of change, and how development priorities evolve within global socio-political realities. The CPWF used water as an entry point to identify the most pressing current and future development challenges within an R4D framework, and solutions to address these challenges.

This approach drew on thinking that links local realities with global influences by understanding how people interact with the complex natural environment. Interlinked planetary boundaries (Rockström et al., 2009) were merged with social boundaries (Raworth, 2012) and overlain with the notion of common-pool resources and collective self-governance (Ostrom, 2009). This provided the framework for the CPWF's R4D that seeks relevance, impact and equity.

The CPWF placed R4D within a context of poverty and development. Poverty has no single definition with measures of poverty ranging from head counts of people living on a certain minimum amount of income to people-centered approaches of how well people meet their livelihood goals. The CPWF focused on people-centered approaches using participatory methodologies while also recognizing the importance of economic dynamics at and between all levels of society. It also included the concept of social exclusion acknowledging that multiple forms of discrimination impact severely on the poor and their capacity to influence decisions that directly affect their lives.

People-centered perspectives allowed the CPWF to consider the causes of poverty, including the importance of human agency, empowerment and institutional accountability. Human agency is what poor people can do for themselves, and empowerment is creating conditions that allow them to do so. These perspectives not only recognize the strategic importance of economic development, but the role of institutions as possible root causes of poverty.

Water and poverty

The CPWF focus on water management and social and ecological resilience[7] led to research on the connections between water and poverty. Water poverty identifies water-specific forms of poverty, such as livelihoods that depend on water, and which are subject to water hazards or lack of development (Black and Hall, 2004; Cook and Gichuki, 2006). For example, people living more than one kilometer from a safe water supply are water poor (Sullivan, 2002).

When the BFPs started in 2005, the CPWF had identified that key issues were the links between water productivity, water scarcity and water poverty.

The question was whether focusing on water could lead to useful insights that could guide interventions. But by the end of the BFPs in 2009, it was clear that water poverty and general poverty were only weakly related. Indeed, "the incidence of poverty and the availability of water are not necessarily linked and severity of poverty depends on the level of control over water, rather than the endowment" (Namara et al., 2010).

The relation between poverty and water across basins was not clear. Shifting the view away from water to the stage of development of the basin, showed that rural poverty was high in underdeveloped basins where agriculture contributed most to total economic output.

Agricultural basins with high levels of rural poverty are characterized by greater use of natural capital than physical capital, and reliance on local, informal institutions rather than the formal state water resources institutions. Industrialized basins had low levels of absolute rural poverty but varying levels of relative poverty. Intermediate basins, which had the greatest total populations in the BFP basins, had pockets of poverty within rapidly changing societies (Kemp-Benedict et al., 2011).

In all basins, water scarcity often had institutional rather than physical causes, but the relevant institutions differed with the basin's place on the development trajectory. In agricultural basins, the dominant institutions are local and traditional, and state institutions are relatively weak. In contrast, in transitional and industrial basins, state institutions dominate. In these basins, rural poverty is concentrated in specific areas that remain poor due to many causes that can only be addressed weakly through technical increases in water productivity. In contrast, in agricultural basins, technical improvement of water productivity can have a substantial impact on poverty.

Interventions that give only modest increases in production, together with reduced variability, may be enough to allow poor farmers in agricultural basins to accumulate assets and diversify their incomes, often outside of water and agriculture. The sustainable livelihoods framework (Box 1.4) is a useful tool to capture modest impacts by combining all of the components of a household's assets both within an institutional context and the larger natural and political environment. Increased financial and human capital can permit diversification and thereby increase resilience.

Box 1.4 The sustainable livelihood framework (SLF)

"In the SLF, households deploy their financial, physical, human, social and natural assets . . . to meet their livelihood goals." "The SLF is a usable way of thinking about development and poverty" (Kemp-Benedict et al., 2011).

Outcome-based R4D and how change occurs

In the CPWF, R4D reflects a shift in understanding of development processes and the role of research. It integrates notions of power and the relationships between people, institutions and partners and their evolving dynamics. It addresses inequities and engages with a diversity of groups and individuals. The relevance of research is transformed and with it the focus, approach and process also change.

R4D for whom?

In Phase 2, the CPWF pursued a path of targeted, inclusive R4D, based on development challenges decided in consultation with partners in six basins (Box 1.5). Scientific research remained a central component, but the research was for transformative change or outcomes. Research for outcomes required understanding of the relevant institutional and social structures. It also implied engagement with partners with the CPWF playing the role of a boundary organization, enabling, linking and translating learning across communities. Effective boundary organizations, which the CPWF aimed to become, depend on their credibility as well as the salience and relevance of the knowledge they share.

Problems can be technical, institutional or political. Problem diagnosis examines the causal relationships among technologies, institutions and policies. It also traces out the nature and value of positive and negative externalities in which the problems being faced by one group are attributable to actions taken by other groups. Water- and food-related problems often involve common property, collective action, property rights and questions of access to resources.

Box 1.5 Basin Development Challenges

Andes basins: To increase water productivity and to reduce water-related conflict through the establishment of equitable benefit-sharing mechanisms.

Ganges: To reduce poverty and strengthen livelihood resilience through improved water governance and management in coastal areas of the Ganges Basin.

Limpopo: To improve smallholder productivity and livelihoods and reduce livelihood risk through integrated water resource management.

Mekong: To reduce poverty and foster development by optimizing the use of water in reservoirs.

Nile: To strengthen rural livelihoods and their resilience through a landscape approach to rainwater management.

Volta: To strengthen integrated management of rainwater and small reservoirs so that they can be used equitably and for multiple purposes.

Theory of change

Initially the CPWF used the CGIAR-wide impact pathways approach, which itself was a shift from the donor-driven logical framework. As the CPWF progressed toward the second phase, theory of change (ToC) (Vogel, 2012) became the dominant conceptual approach. "[T]heory of change represents people's understanding of how change happens—the pathways, factors and relationships that bring and sustain change in a particular context" (James, 2011).

Although ToC was the main conceptual approach, the CPWF used other frameworks for differing specific purposes, when the alternative approach was judged more suitable. For example, as discussed above, the poverty and livelihoods analysis used the sustainable livelihoods framework (Solesbury, 2003; Kemp-Benedict et al., 2009), while the political economy analysis used the institutional analysis and development framework (Harris et al., 2011).

In developing ToCs at the project, basin and program level in CPWF's second phase, the wide diversity of people involved in the range of CPWF R4D (partner research organizations, local decision-makers, policymakers, development practitioners, etc.) themselves contributed to defined development outcomes. The CPWF model of practice approached R4D through ToC thinking. It put ToC into practice using a set of tools, such as outcome learning models, regular reflection meetings and use of "most significant change" stories, all of which were developed iteratively.

The CPWF's experience demonstrated the value of the ToC approach. ToC created narratives that were accessible to all participants. These narratives were established through a combination of collective inclusive reflection, adaptive management and relating change to specific groups of actors. ToC also recognized that explicitly stated assumptions are often subjective and depend on people's cultural and socio-economic perspectives.

> Every programme is packed with beliefs, assumptions and hypotheses about how change happens—about the way humans work, or organisations, or political systems, or eco-systems. [ToC] is about articulating these many underlying assumptions about how change will happen in a programme.
>
> (Rogers, 2008)

Achieving outcomes with information and engagement

The CPWF defined outcomes as changes in practice, in behavior, decisions, investments or other ways in which people choose to do things differently. This is not coercing people to do things differently, but engaging with them to help them obtain information that allows them to make informed choices because they perceive the change to be to their own advantage. R4D therefore seeks to contribute to development outcomes that are profitable, equitable, sustainable and resilient. The CPWF used ToC to describe the process, which reflects an inclusive, participative and reflective learning process.

Understanding the process of engagement is crucial (Box 1.6). Engagement is also part of problem definition in which the CPWF encouraged stakeholders to participate to achieve a common vision of the nature of the problem, its causes and drivers, and what might be done about it.

In R4D, the CPWF distinguished between research to define development issues and research to identify feasible and socially acceptable solutions (also called interventions, strategies, etc.). Research for solutions required sound understanding of the issues for which a solution is sought, including taking account of the scale (region, basin, catchment, etc.).

Effective solutions are often those that integrate improved technologies, new institutional arrangements and reformed policies, all three of which may co-evolve. Research on solutions may find win-win strategies to overcome contentious issues, or may define trade-offs to support negotiations. They may also be site specific, the conditions of which must be defined as part of targeting. They will generate a range of consequences on profits, livelihoods, gender equity, downstream resource users, ecosystem services, resilience, and so on, some of which may be unexpected. Research on solutions therefore needs to be dynamic and inclusive to respond to whatever may arise.

Research on solutions must also be sensitive to the policy environment, align where it is appropriate and maintain its relevance. In some cases policies may

---

Box 1.6 The CPWF experience with engagement

Engagement is most effective when it:

- is evidence-based, well informed by research products, and builds on long-term relationships by working through existing networks (instead of creating new ones);
- understands power relations by bringing in people with authority and responsibility for taking major decisions;
- recognizes as honest brokers groups that have different and conflicting interests;
- fosters negotiation when dealing with management of common property;
- continues for a long time, often for a series of outcomes, which collectively enhance impact;
- generates key messages tailored to different stakeholders;
- enables all partners to understand and address the problem participatively; and
- identifies and develops credible champions with vision of what can be achieved and who are involved in the long term.

Authors discuss engagement in more detail in Chapters 3 and 5.

obstruct the use of attractive solutions, while in others favorable policies can be leveraged to make fast progress. In all cases understanding the policy environment and how to impact it is crucial.

The CPWF found that it was often more effective to be an influential, credible and respected member of a third party's network rather than create one for itself. Moreover, both the process of defining problems and then discovering solutions to them, and the process of engagement took place at the same regional or basin level.

Conclusions

The CPWF started in 2002 with the objective to "significantly increase the productivity of water used for agriculture . . . in a manner that is environmentally sustainable and socially acceptable." We have shown how this evolved from producing the outputs of conventional science to a R4D approach that used water-related innovations to involve partners in all stages of the process to produce outcomes. It carried out three functions:

- Better understand and define water-related problems and challenges at different scales (Chapter 2).
- Better understand the intricacies of designing water-related innovations and understanding their performance under different conditions, as well as their consequences for livelihoods, equity and the environment (Chapters 3, 4 and 5).
- Better understand how to engage with stakeholders to foster dialogue and negotiations to lead to equitable development outcomes (Chapters 6 and 7).

These three functions comprise a widened notion of development and change in which research plays a role in defining development pathways. As the CPWF progressed, the research process changed, results became focused on development outcomes, contribution to impact at scale became feasible, and a range of tools, approaches and frameworks complemented each other. In the chapters that follow authors describe the process and outcomes in more detail.

Notes

1   The abbreviation CGIAR was for the Consultative Group on International Agricultural Research from 1971 to 2010. The institution was restructured in 2010, incorporating the abbreviation as part of the name of the new entity, the CGIAR Consortium of International Agricultural Research Centers.
2   TAC document SDR/TAC:IAR/00/14.1/Rev.2.
3   sciencecouncil.cgiar.org/fileadmin/templates/ispc/documents/Publications/ 1a-Publications_Reports_briefs_ISPC/TAC_Food-Secure-World-for-All_2000.pdf (accessed 8 April 2014)
4   The HarvestPlus Challenge Program to produce bio-fortified crops, coordinated by CIAT, and the Generation Challenge Program to use advanced genetic technologies to improve crops for greater food security in the developing world, coordinated by

Centro Internacional de Mejoramiento de Maíz y Trigo [International Maize and Wheat Improvement Center) (CIMMYT) and International Rice Research Institute (IRRI).
5 Minutes of the 82nd meeting of the iSC at Centro Internacional de la Papa [International Potato Center](CIP), 8–12 April, 2002. Available from: library.cgiar. org/bitstream/handle/10947/5684/iscchairreport.pdf (accessed 8 April 2014).
6 *External Review of the Challenge Program on Water and Food.* Available from: gppi.net/fileadmin/gppi/Markus_exco13_cpwf_cper.pdf (accessed 8 April 2014).
7 "A key concept in the resilience framework is the concept of social-ecological systems. There are no natural systems without people, nor social systems without nature. Social and ecological systems are truly interdependent and constantly co-evolving" (Stockholm Resilience Centre, 2007).

References

Black, M. and Hall, A. (2004) 'Pro-poor water governance', in: *Water and poverty: The themes.* Asian Development Bank, adb.org/sites/default/files/pub/2004/Themes_04.pdf, pp. 11–20 (accessed 8 April 2014).
Cook, S. and Gichuki, F. (2006) *Analyzing water poverty: Water, agriculture and poverty in basins,* CPWF BFP Working Paper No. 3, CGIAR Challenge Program on Water and Food, Colombo, Sri Lanka.
Cosgrove, W. J. and Rijsberman, F. R. (1998) 'Creating a vision for water, life and the environment', *Water Policy*, vol. 1, pp. 115–122.
Deloitte (2012) *Water tight 2012.* deloitte.com/assets/Dcom-SouthAfrica/Local%20Assets/Documents/water_tight_2012.pdf (accessed 8 April 2014).
Harris, D., Kooy, M. and Jones, L. (2011) *Analysing the governance and political economy of water and sanitation service delivery,* Working Paper 334, Overseas Development Institute, London, odi.org.uk/resources/docs/7243.pdf (accessed 8 April 2014).
James, C. (2011) *Theory of change review: A report commissioned by Comic Relief,* Comic Relief, London, mande.co.uk/blog/wp-content/uploads/2012/03/2012-Comic-Relief-Theory-of-Change-Review-FINAL.pdf (accessed 8 April 2014).
Kemp-Benedict, E., Bharwani, S., de la Rosa, E., Krittasudthacheewa, C. and Matin, N. (2009) *Assessing water-related poverty using the sustainable livelihoods framework,* SEI Working Paper, Stockholm Environment Institute, Stockholm.
Kemp-Benedict, E., Cook, S., Allen, S. L., Vosti, S., Lemoalle, J., Giordano, M., Ward, J. and Kaczan, D. (2011) 'Connections between poverty, water and agriculture: Evidence from 10 river basins', *Water International*, vol. 36, no. 1, pp. 125–140.
KPMG (2012) *Water scarcity: A dive into global reporting trends,* KPMG Sustainability Insight 2012, kpmg.com/Global/en/IssuesAndInsights/ArticlesPublications/sustainable-insight/Documents/sustainable-insights-water-survey.pdf (accessed 8 April 2014).
McKinsey and Company (2012) *Resource revolution: Meeting the world's energy, materials, food, and water needs,* mckinsey.com/features/~/media/mckinsey/dotcom/home page/2011%20nov%20resource%20revolution/resource_revolution_full_report_v2.ashx (accessed 8 April 2014).
Namara, R. E., Hanjra, M. A., Castillo, G. E., Ravnborg, H. M., Smith, L. and Van Koppen, B. (2010) 'Agricultural water management and poverty linkages', *Agricultural Water Management*, vol. 97, pp. 520–527.
Ostrom, E. (2009) 'A general framework for analyzing sustainability of social-ecological systems', *Science*, vol. 325, no. 5939, pp. 419–422.

Raworth, K. (2012) *A safe and just operating space for humanity: Can we live within the doughnut?*, Oxfam Discussion Papers, Oxford, UK, oxfam.org/sites/www.oxfam.org/files/dp-a-safe-and-just-space-for-humanity-130212-en.pdf (accessed 8 April 2014).

Rijsberman, F. R. (2004) 'Water scarcity: Fact or fiction?', in: T. Fischer, N. Turner, J. Angus, L. McIntyre, M. Robertson, A. Borrel and D. Lloyd (eds) *New directions for a diverse planet*. Proceedings of the 4th International Crop Science Congress, 26 September–1 October 2004, Brisbane, Australia, cropscience.org.au/icsc2004/plenary/1/1994_rijsbermanf.htm (accessed 8 April 2014).

Rockström, J., Steffen, W., Noone, K., Persson, Å., Chapin, F. S., Lambin, E. F., Lenton, T. M., Scheffer, M., Folke, C., Schellnhuber, H. J., Nykvist, B., de Wit, C. A., Hughes, T., van der Leeuw, S., Rodhe, H., Sörlin, S., Snyder, P. K., Costanza, R., Svedin, U., Falkenmark, M., Karlberg, L., Corell, R. W., Fabry, V. J., Hansen, J., Walker, B., Liverman, D., Richardson, K., Crutzen, P. and Foley, J. A. (2009) 'A safe operating space for humanity', *Nature*, vol. 461, no. 7263, pp. 472–475.

Rogers, P. J. (2008) 'Using programme theory for complicated and complex programmes', *Evaluation*, vol. 14, no. 1, pp. 29–48.

Sayer, J. A. and Campbell, B. M. (2003) 'Research to integrate productivity enhancement, environmental protection, and human development', in: B. M. Campbell and J. A. Sayer (eds) *Integrated natural resource management: Linking productivity, the environment and development*, CABI, Wallingford, UK, pp. 1–14.

Solesbury, W. (2003) *Sustainable livelihoods, a case study of the evolution of DFID policy*, Working Paper 217, Overseas Development Institute, London, odi.org.uk/resources/docs/172.pdf (accessed 8 April 2014).

Stockholm Resilience Centre (2007) 'Social-ecological systems', available from: stockholmresilience.org/21/research/what-is-resilience/research-background/research-framework/social-ecological-systems.html (accessed 8 April 2014).

Sullivan, C. (2002) 'Calculating a water poverty index', *World Development*, vol. 30, pp. 1195–1210.

The 2030 Water Resources Group (2009) *Charting our water future*, 2030water resourcesgroup.com/water_full/Charting_Our_Water_Future_Final.pdf (accessed 8 April 2014)

Vogel, I. (2012) *Review of the use of 'Theory of Change' in international development: Review report*, dfid.gov.uk/r4d/pdf/outputs/mis_spc/DFID_ToC_Review_VogelV7.pdf (accessed 8 April 2014).

World Commission on Environment and Development (1987) *Our common future*, Oxford University Press, Oxford.

2 Water scarcity and abundance, water productivity and their relation to poverty

Alain Vidal,[a][*] *Larry W. Harrington*[b] and *Myles J. Fisher*[c]

[a]CGIAR Challenge Program on Water and Food CPWF, Montpellier, France; [b]CGIAR Challenge Program on Water and Food CPWF, Ithaca, NY, USA; [c]Centro Internacional de Agricultura Tropical CIAT, Cali, Colombia; [*]Corresponding author, a.vidal@cgiar.org.

Water scarcity and beyond

The Challenge Program on Water and Food (CPWF) was conceived as a response by the Consultative Group on International Agricultural Research (CGIAR) to a perceived global crisis: the threat posed by water scarcity to food security, livelihoods and the environment, and the urgent need to use increasingly scarce water resources more efficiently. With the passage of time, the CPWF has broadened its agenda to focus on a range of development challenges in basins that relate to water. The CPWF came to see that water provides a useful entry point for addressing many development challenges, including those related to sustainable intensification of agricultural systems and preservation of ecosystem services. Addressing water scarcity is a means to a broader end as well as an end in itself.

In this chapter, we look back at some of the concepts that underpinned the original CPWF. We review recent findings on water scarcity at the global level and compare these with basin-level information on water scarcity from CPWF Basin Focal Projects (BFPs). We also take a closer look at the multiple dimensions of water scarcity as they affect farm family livelihoods and show that water can be scarce even when it is apparently abundant. We then revisit the concept of water productivity (WP) (embodied in the phrase, "crop per drop") and discuss its usefulness and limitations as an indicator. Finally, we review what the CPWF has learned regarding the subtle and complex relationships among water scarcity, poverty, livelihoods and food security.

The global level—freshwater is scarce

The essence of the global water scarcity narrative is simple: freshwater supply and demand are out of balance in important regions and the mismatch is likely to get worse. The narrative suggests that the demand for water-related products (especially food) will grow faster than the population increases, whereas the supply of freshwater is limited, and that the main question is the timing and spatial incidence of the imbalance between demand and supply. We do not entirely concur with this narrative. Water scarcity means different things to different people in different environments. Water can be both abundant and scarce in the same environment, depending on the kinds of water and water uses discussed.

Some observers are blunt: "Water shortages have emerged as one of the most important infrastructure issues in the world today . . . Global demand for freshwater will exceed supply by 40% by 2030 . . . with potentially calamitous implications for business, society and the environment" (KPMG, 2012). Recent reports speak of "water bankruptcy" for many regions (Mee and Adeel, 2012) while water shortage has been called "the defining crisis of the 21st century" (Pearce, 2007).

Scenario analysis used in the World Water Vision for 2000 warned that continued "business as usual" water management was likely to result in, "a global system . . . becoming more and more vulnerable as a result of the increasing scarcity of water resources per capita, the diminished quality of water and increasing conflicts associated with inequality, water scarcity, and the narrower resource base of healthy ecosystems." Scenario analysis took account of nearly two dozen drivers of change, among them demographic, economic, technological, social, governance, and environmental [hydrological] factors (Gallopín and Rijsberman, 2000).

Increased demand for food (population growth and dietary changes), rapidly growing megacities, urbanization and industrialization, biofuel production, and the increasing effects of climate change all drive increased use of freshwater. Demand for drinking water and sanitation services will be a factor, but the real increase in water demand will come from agriculture to produce food, feed, fiber and fuel. With human populations predicted to increase from 7 billion in 2012 to about 9 billion in 2050, agricultural water requirements may grow to as much as 14,000 billion m^3/yr (Chartres and Varma, 2011), or almost double (see below). These predictions are based on current levels of agricultural WP, including rainfed as well as irrigated agriculture.

These scenarios of water demand are only slightly higher than those published in the *Comprehensive Assessment of Water in Agriculture* (CA), which noted that, "without further improvements in water productivity or major shifts in production patterns, the amount of water consumed by evapotranspiration in agriculture will increase by 70%–90% by 2050. The total amount of water evaporated in crop production would amount to 12,000–13,500 [billion m^3/yr], almost doubling the 7130 [billion m^3/yr] of today" (Molden, 2007).

Other analyses estimate annual water use in agriculture at 8500–11,000 billion m^3/yr by 2050 (Rockström et al., 2010). They assume some growth in productivity and separate consumption in rainfed (6500–8500 billion m^3/yr) from irrigated (2000–2500 billion m^3/yr) agricultural systems.

The above estimates focus on demand for agricultural water, but ignore the supply side. How bad is the mismatch between demand and supply of agricultural water? We discuss three ways to address this question—through analysis of planetary boundaries; threats to water security from multiple stressors at sub-national levels; and economic versus physical water scarcity in basins.

Analysis of planetary boundaries defines boundaries, "within which we expect that humanity can operate safely. Transgressing one or more planetary boundaries may be deleterious or even catastrophic due to the risk of crossing thresholds that will trigger non-linear, abrupt environmental change within continental- to planetary-scale systems" (Rockström et al., 2009). These boundaries include climate change, ocean acidification, stratospheric ozone, global P and N cycles, atmospheric aerosol loading, land use change, biodiversity loss, chemical pollution, and use of freshwater. Recent analysis concludes that passing a boundary of about 4000 billion m^3/yr of consumptive use of blue water[1] will increase the risk of collapse of terrestrial and aquatic ecosystems (Rockström et al., 2009). The analysis focuses on blue water, however, it is not restricted to agricultural water nor does it address the question of whether water use in agriculture will substitute for natural land use.

An analysis of threats to water security from multiple stressors at sub-national levels takes a different slant (Vörösmarty et al., 2010). It uses spatial accounting to assess threats to human water security, where a threat is exposure to stressors at given location. There are four categories of stressors: catchment disturbance, pollution, water resource development and biotic factors. Catchment disturbance includes cropland use, impervious surfaces, livestock density, wetland and disconnectivity. Pollution includes such factors as soil salinization, loading of excess plant nutrients, toxic materials and sediments, acidification and thermal alteration. The analysis concludes that 80 percent of the global population is exposed to high levels of threat.

Areas not exposed include parts of the Amazon, central Africa, the Malay Archipelago, and parts of southeast China and Southeast Asia with low populations and high rainfall. Rich countries make massive investment to offset high stressor levels. All CPWF basins are located in areas with high levels of threat to water security.[2] The analysis focuses on blue water in rivers; however, it is not restricted to agricultural water and emphasizes quality more than availability.

The basin level—blue water is (sometimes) scarce

Although much rainwater is unused by people, water scarcity is an important topic. With regard to blue water, the CA (Molden, 2007) distinguished between areas with no water scarcity, with economic water scarcity, and with physical water scarcity (Rijsberman, 2006).

Physical water scarcity occurs when withdrawal of water approaches or exceeds sustainable limits commonly set at 75 percent of the river flow. This may be because of a lack of supply, high demand, or both. Economic water scarcity occurs when there is inadequate investment in water-related infrastructure, which limits access to water even where there is no local physical scarcity and withdrawal is less than 25 percent (Molden, 2007). High levels of water use lead to closed basins, that is, where water no longer flows out through the rivers. The Yellow River in China failed to reach the sea in 1997 for 226 days and was dry to 600 km upstream (Ringler et al., 2012). Subsequent government action reduced water use for irrigation and this helped provide year-round flows, which, however, were still insufficient to counter entry of sea water within the basin, leading to damage of wetlands. The Karkheh in Iran is potentially closed by the use of its limited water for irrigation downstream of the new dam to the detriment of the Hoor-al-Azim wetlands on the border with Iraq (Ahmad and Giordano, 2012).

By these definitions, only two of the CPWF's ten basins, the Limpopo and the Yellow, suffer from physical water scarcity, although parts of the Ganges and the Karkheh have scarcity at times in specific places.

Both population size and availability of blue water affect water scarcity, and both are captured by the Falkenmark water stress indicator (Falkenmark, 1997) and can be applied to basins, countries or regions. *Water stress* occurs when there are less than 1700 m³/yr of renewable water resources per capita for all needs. When per capita supply falls below 1000 m³/yr there is *water scarcity*, and below 500 m³/yr, *absolute scarcity*. The indicator has great spatial variability, across continents, within river basins, across and within countries. It falls faster where population growth is rapid. The Yellow (1250) is stressed, the Karkheh (1970) and the Volta (2560) approach stress, but even the populous Indus-Ganges (5900) is over three times the level that indicates stress. The other basins exceed 10,000 (the São Francisco is the highest at 38,390). In the future, many more people are likely to experience water stress and water scarcity, especially in sub-Saharan Africa and South Asia (Figure 2.1) (UNDP, 2006).

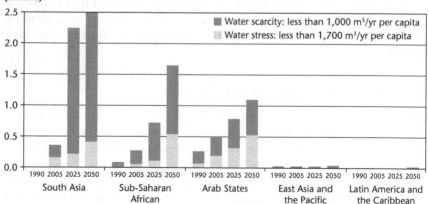

Figure 2.1 Water stress is projected to accelerate in intensity in several regions.
Source: UNDP, 2006.

Taken together, these analyses suggest that blue water is only physically scarce in selected areas. Problems with water quality and lack of investment in water storage technology are far more pervasive than scarcity. None of the analyses consider green water. We conclude that institutions and governance are central to dealing with issues of water control, water quality and infrastructure investment.

The basin level—green water is often abundant

At the level of a river basin, freshwater can be scarce and abundant at the same time, depending on whether we are talking about blue water or green water. Green water and rainfall may be abundant when blue water is scarce. Even in dry basins, green water can be deemed abundant when only a small proportion of precipitation or actual evapotranspiration (AET) goes through agriculture.

Blue water makes up rivers, lakes and groundwater. Green water includes water stored in the soil to be transpired in the process of vegetative productivity. There is a complicated interaction between green and blue water involving precipitation, temperature, topography, soil type, vegetation cover and processes that control runoff and deep drainage (Chartres and Varma, 2011). Analysis of future demand and supply for freshwater for the most part focuses on blue water, but agriculture uses three to four times more green than blue water.

Evapotranspiration (ET) is the sum of plant transpiration and evaporation from soils and open water to the atmosphere. Meteorologists differentiate between potential ET (PET), which is the atmosphere's ability to evaporate water, and AET, which is the amount of water that does evaporate from all sources. Agricultural uses, including pasture, often account for only a small proportion of AET, so that a lot of green water does not go through agriculture at all (Table 2.1). In the moderately dry Volta Basin, for example (average precipitation 973 mm/yr), only about 10 percent of precipitation goes through agricultural systems, accounting for 11 percent of AET.

More productive use of rainwater can therefore help to resolve the global crisis of freshwater scarcity. We need to focus on the blue–green water nexus (Falkenmark and Rockström, 2010), that is, on green water as well as on blue water since transpiration from vegetation is a major water use.

The BFPs

The CPWF BFPs research provided information on the distribution of water across different environments and land uses. From 2005 to 2009, the BFPs researched water availability, water balances, WP, the relationships between water, poverty, and other factors in ten river basins.[3] The BFP basins cover a wide range of geographic settings on three continents with considerable cross-basin and within-basin variability in size, topography, land use, extent of irrigation, population density, income levels, poverty, precipitation, temperature,

Table 2.1 Agricultural use of AET and rainfall.

	Annual total rainfall mm/yr	Mean AET mm/yr	Mean AET for productive pastures and agricultural areas mm/yr	Agricultural use of AET %	Agricultural use of rainfall %
Andes	784	632	43	7	5
Ganges	1073	746	499	67	47
Karkheh	348	291	13	4	4
Limpopo	547	640	103	16	19
Mekong	1713	1049	393	37	23
Niger	1017	804	116	14	11
Nile	618	606	36	6	6
São Francisco	975	928	94	10	10
Volta	973	910	98	11	10
Yellow	438	458	229	50	52

Source: Mulligan et al. (2012a).

seasonality, water resources infrastructure, groundwater resources, water access for direct consumption or for agriculture, PET and AET, agricultural water demands, domestic and industrial use (Mulligan et al., 2012b).

A series of water use accounts gave details of catchment-lumped water availability and water balances in all ten basins (e.g. Kirby et al., 2010b; Eastham et al., 2010). Kirby et al. (2010a) discussed the methods used. The CPWF published the principal research outputs of the BFPs and syntheses of the various components in *Water, Food and Poverty in River Basins: Defining the Limits* (Fisher and Cook, 2012). We shall draw further on the BFP research in several of the following chapters.

The multiple dimensions of farm-level water scarcity

Problem definition

Global- and basin-level estimates of present and future water supply and demand are important to establish the limits for water allocation and use by helping to define problems of water scarcity and WP. However they gloss over the many complex ways in which water scarcity affects farm productivity and family livelihoods.

In this section, we focus on the components of water scarcity and their effects on how families manage their farm systems. In this context, we define water scarcity as a *failure to achieve the right amount of the right quality of water for the right purpose at the right time for the right people*.

To make this definition operational, we need to define what we mean by "right" in each context, understanding that its meaning depends on whose viewpoint we are representing. In the end, water is scarce for someone in some way nearly everywhere.

The right amount means that many crops have specific needs: while paddy rice needs to be flooded, in contrast many crops are sensitive to waterlogging. The purpose that water will be used for determines what the right quality is, for example, some crops can tolerate more salinity than others so that the salinity of irrigation water determines which crops can be grown with it. Other aspects of quality include maintaining sediments and other pollutants at acceptable levels. We can define the right purpose either narrowly in terms of crop, agricultural system or landscape management, or broadly in terms of water allocation across a wide range of ecosystem services. We use the right time to take account of seasonality, and how seasonal patterns change over time. The right people means equitable allocation between alternative groups of water users, including people who benefit from water-related ecosystem services, and sometimes water managers or polluters.

Several of these factors often come together to create scarcity where water seems abundant. Water appears plentiful in coastal Bangladesh for most of the year, but it is a water-scarce environment for some purposes. There is not enough water of the right quality (salinity < 2 g/L) available at the right time (end of wet season and throughout the dry season) for the specific purpose of completing wet-season rice crops, followed by dry-season crops/aquaculture. When the wet season ends with decreasing river flows, sea water intrusion affects the quality of the river water surrounding the polders, so that quality, timing and purpose together create scarcity (PN10[4]) (Tuong and Hoanh, 2009).

Rainfall in the Ethiopian highlands is at least 1300 mm/yr and ET is modest, yet water is often scarce for crops and livestock. The causes are sloping landscapes that give high rates of runoff, soils with low water-holding capacity and poor infrastructure for water-storage (Block, 2008; Awulachew et al., 2010).

We now discuss water scarcity in terms of aridity, seasonal unreliability, quality, excess and access using examples from CPWF projects.

Aridity

The water scarcity of arid lands, which have no rainy season, is physical scarcity caused by low precipitation and high atmospheric demand. The ratio of mean annual precipitation to PET—the aridity index (Middleton and Thomas, 1997)—ranges from zero to less than 0.20 in those regions. Water quality and allocation are not part of this index.

Arid areas can only sustain agriculture with irrigation. Elsewhere they are rangelands used at low intensity for ruminant production, often with nomadic herders ranging 1000 km or more in their yearly transhumance. The herders require access to crop residues and watering points, which is a critical component that is coming under threat in the northern Sahel (PN64) (Clanet and Ogilvie, 2009). The rangelands of the dry, central Limpopo are also grazed at low intensity but the pastoralists are sedentary (PN62) (Sullivan and Sibanda, 2012).

It is important to distinguish aridity from "physical water scarcity" as defined in the CA ("when more than 75% of the river flows are withdrawn for agriculture, industry and domestic purposes") (Molden, 2007). The former focuses on rainfall, the latter on the extent of blue water withdrawals.

The CPWF had few projects in catchments or sub-basins in arid areas. Most projects were located where other dimensions of water scarcity were more important.

Seasonal unreliability

Outside of arid regions, average annual rainfall is adequate for agriculture of some kind, but annual averages can conceal more than they reveal. Rainfall may fail when it is most needed, and the more unreliable the rainfall, the more frequent failures. Unreliable rainfall can reduce productivity and favor extensive use of land to reduce the risk, for example low planting densities. Unreliability may affect contrasting social groups differently and with varying levels of severity.

Unreliability may be normal or exceptional. Where it is normal, farm families are likely to have developed multi-layered mechanisms to cope. If unreliability becomes extreme, coping mechanisms may fail and threaten family survival. Seasonal unreliability of rainfall is only part of the story. Risk of loss is higher when farm families lack coping mechanisms and when investment in water infrastructure and management is inadequate. The same problems of seasonal unreliability may affect different social groups in different ways.

There are several dimensions to the problem of seasonal unreliability, some of them closely related:

- Seasonality of the rainy season (one or more of late onset, early termination, extended dry periods within the season, unfavorable temporal distribution or outright failure);
- Seasonality of the supply of stored water (inadequate quantity of stored water to grow crops or fodder in the dry season);
- Seasonality of the demand for stored water (the demand increases in years when the rainy season is short, unreliable, or when failure of rainy season crops due to pests or disease forces farmers to resow);
- Seasonality of water quality (excess of saline water when freshwater is needed [for rice] or excess of freshwater when saline water is needed [for shrimp]);
- Seasonality of river flow and flooding (inadequate or excessive pulsing of river systems to support catch fisheries or aquaculture; unanticipated and excessive seasonal flooding that destroys crops and livestock). Many

farming systems and capture fisheries in water-rich basins such as the Mekong are adapted to seasonal flooding. According to the timing and extent of the floods, however, they may destroy wet season crops, or they may enable wet- or dry-season cropping.

The BFPs assembled basic information on annual rainfall, and its seasonality as measured by the coefficient of variation of monthly rainfall (Table 2.2).

Unreliability can contribute to poverty traps as noted by Grey and Sadoff (2002, p. 4) regarding Africa:

> We have all witnessed ... catastrophic flood and drought—the endemic and unpredictable consequence of Africa's hydrological variability. The economic impacts can be a significant proportion of GDP and social impacts are incalculable [as is] the suffering of individual families and communities, as years of labor in land preparation and crop development is withered by drought or washed away by flood ... the very existence of extreme variability itself creates disincentives for investment and affects the performance and structure of economies, as the unpredictability of rainfall and runoff encourages risk averse behavior in all years, promoting patterns of development that can trap economies in a low-level equilibrium. Thus, even in years of good rains, economic productivity and economic development can be constrained by conditions of hydrological variability.

In the Limpopo Basin, 80 percent of the annual precipitation falls between November and late February with a mean of 50 rainy days. Variability in rainfall, soil type, ground cover, and slope gives erratic runoff and pronounced seasonal variation in flow, with negligible flow in the dry season. Seasonal rainfall patterns vary unpredictably and substantially from one year to the next. (PN62) (Sullivan and Sibanda, 2012). The Volta Basin has more rainfall than

Table 2.2 Annual precipitation and its seasonality in the BFP basins.

Basin	Annual total rainfall mm/yr	Precipitation seasonality CoV%
Andes	784	78
Ganges	1073	125
Karkheh	348	89
Limpopo	547	84
Mekong	1713	86
Niger	1017	108
Nile	618	103
São Francisco	975	84
Volta	973	96
Yellow	438	93

Source: Mulligan et al. (2012a).

the Limpopo, but seasonal unreliability of rainfall affects it almost as much. Rainfed agriculture only uses 14 percent of the basin's rainfall, but drought years and within-year dry spells, together with the infertility and low water-holding-capacity of the soils, cause low crop yields and WP (PN55) (Lemoalle, 2008). Even in the high-rainfall highlands of the Nile Basin, drought and the intra-seasonal variability of rainfall causes crop failures, livestock deaths and livelihood disasters (Nile 2) (Amede et al., 2007).

Seasonal unreliability adversely affected many CPWF projects. A project in the Limpopo Basin noted that, "Rainfed smallholder cropping in semi-arid Zimbabwe is constrained by frequent droughts and mid-season dry spells . . . In southern Zimbabwe, it is actually rare for drought or mid-season dry spells not to occur and this has led to permanent food insecurity for the majority of households" (PN17) (Mupangwa et al., 2011). In another Limpopo project, rainfall was so erratic that researchers could not establish cropping trials or the trials failed with no grain harvest. In drier years, structures to harvest rainwater were ineffective because there was not enough rainfall to collect, while in wetter years they were often washed out (PN1) (Siambi, 2011).

Rainfall variability causes risk and uncertainty. Early sowings can fail if there is early-season drought, while late-season drought or competition from early-season weed growth can reduce the yields of late sowings. In either case, mid-season dry spells can further reduce yields. Farmers mostly know the risks of unreliable rainfall and use many strategies to manage it (Scoones, 1996; Harrington and Tow, 2011) (Box 2.1).

Unreliable rainfall constrains the use of fertilizer and other inputs when the risk of crop failure outweighs their potential benefits (CIMMYT, 1999). Under some conditions, however, fertilizer micro-dosing or applying low levels of

---

Box 2.1 Farmers' strategies to manage risk of seasonally unreliable rainfall

Staggered planting dates;
early-maturing varieties;
varieties with different crop durations;
crop combinations (for example both maize and sorghum or millet);
dry-season plowing to control weeds and allow earlier sowing;
reduced planting density;
intercropping;
matching crop species to land niches;
supplementary irrigation;
rainwater harvesting; and
seasonal use of wetlands.

(PN55) (Cooper et al., 2008; Terrasson et al., 2009)

basal fertilizer can reduce risk (PN1) (Dimes et al., 2005) when combined with soil cover or cover crops (FUNDESOT, 2012). Pastoralists and agro-pastoralists have developed various community-level coping mechanisms in response to seasonal unreliability.

There are places in China where the rainfall can be too little for flooded rice in some years, but in other years, there is too much rain for maize or other rainfed crops. A CPWF project selected and mapped these places and showed that aerobic rice can grow well when rainfed, but it is not affected by flooding (PN16) (Bouman, 2008). The challenge will be whether aerobic rice will work elsewhere (Rubiano and Soto, 2008).

Quality

We also addressed water quality as a component of scarcity, especially when linked to seasonality ("right amount of the right quality of water for the right purpose at the right time").

Salinity induces seasonal scarcity of freshwater in places such as coastal Bangladesh,[5] where a series of polders create areas of land protected from river flooding or seawater incursion by embankments. Freshwater surrounds the polders during the wet season and salt water during the dry season. Lack of freshwater at the end of the wet season and during the dry season hinders intensification and diversification of farms in the polders. Most produce only one low-yielding rice crop during the wet-season each year.

Nevertheless, in places in coastal Bangladesh it is possible to grow two rice crops in the wet season plus a dry-season crop, or rice followed by aquaculture. The intensification depends on allowing freshwater to enter and be stored when the water surrounding the polders is fresh, and closing the sluice gates as the water becomes saline (PN10 and G2) (Sharifullah et al., 2008; Humphreys, 2012). Overcoming water-scarcity problems caused by variable quality during the year depended on new crop-management technology and new institutions to coordinate management of sluice gates and infrastructure within the polders (G3) (Mukherji, 2012).

Salt stress is a problem in rice in the lower Ganges Basin of India and in Bangladesh without polders. Seawater intrusion causes salinity in coastal areas and inland there are shallow, saline water tables. In the wet season, flooding is a problem, while salinity damages crops during the dry season, and in inland areas, it is expanding. Project PN07 integrated salt-tolerant rice and other crops with complementary land and water management to minimize the effects of salt (Srivastava et al., 2006; Castillo et al., 2007; Vadez et al., 2007; Islam et al., 2008; Ismail, 2009).

In the Andes, sediment often reduces water quality, causing scarcity downstream because muddy water is unsuitable for sprinkler or drip irrigation or for domestic use without expensive treatment. Projects PN22, Andes 2 and Andes 3 promoted institutional changes that allow for payment for ecosystem services and other benefit-sharing mechanisms to encourage farmers in upper catchments to manage their land and water better. They identified hotspots of erosion, measured their impacts on water quality and identified land management that reduces erosion and so improves downstream water quality (Estrada et al., 2009; Quintero et al., 2009; Quintero, 2012).

Throughout the Andes, mining is notorious for contaminating water. It is a growing cause of scarcity of clean water and was researched in the Andes BFP project (Mulligan et al., 2009; Mulligan et al, 2012b). In the *Conversatorio de Acción Ciudadana* process in Colombia, communities and institutions negotiated legal agreements related to water in catchments (Candelo et al., 2008), which inter alia find ways for benefits from mining to be used to address its negative externalities. For example, benefit-sharing mechanisms have been negotiated and implemented in ways that recognize the negative impacts of mining on water and provide the resources to manage water quality directly at the mine to reduce these impacts or by supporting improved management of other land uses. Because mining is an important source of income, both locally and nationally, it requires institutional tradeoffs when the national priority is to reduce poverty (Johnson et al., 2009).

In the Volta Basin, muddy water in the wet season causes scarcity. Rainfall is not scarce in the basin as a whole but it is seasonal and varies from 1200 mm/yr in the south to less than 500 mm/yr in the north. Wet-season runoff is difficult to store because much of the basin is too flat to build large dams (PN55) (Lemoalle and de Condappa, 2012), but several thousand small dams built over the last 20 years supply water in the dry season for domestic use, livestock and small-scale irrigation. The small dams are in streams that are hydrologically linked (PN46) (Andreini et al., 2010).

Most small dams in the White Volta Basin have problems with water quality caused by cyanobacteria (potentially harmful microalgae) of unknown origin. Pesticides and other pollution from agriculture also reduce water quality (PN46) (Andreini et al., 2010). Cyanobacteria constrain the use of water from small dams for households, fishing or irrigation (V3). Small dams also increase the incidence of schistosomiasis and malaria (Boelee et al., 2009). Nevertheless, water quality is better when communities improve their soil management and use of pesticides (V3) (Cecchi and Sanogo, 2012).

In the Nile, water quality differs between upstream and downstream. Siltation and livestock-related water pollution affects water quality in upstream countries, leading to sedimentation of reservoirs and low quality for domestic water. In downstream countries, ET is high, increasing salinity of the river water, which in the delta reduces yields and limits the range of crops that farmers can grow.

Another example of scarcity of water of suitable quality for particular purposes comes from Ghana, where urban and peri-urban vegetable farmers use urban wastewater for irrigation, posing a public-health risk. A CPWF project developed strategies to safeguard public health without compromising farmers' livelihoods. It assessed land and WP in farms irrigating with wastewater and quantified levels of contamination on vegetables at points down the food chain. It then identified low-cost strategies to reduce the risk, which tests by farmers and consumers showed to work. The project's success influenced policy in Ghana to allow the use of urban wastewater (PN38) (Abaidoo et al., 2009; CPWF, 2012).

Excess

Excess water fits our definition of "not being the right amount". It can vary from the brief aftermath of a rainstorm, which may damage crops sensitive to waterlogging, to massive flooding. Floods can result in:

* ruinous damage to farms and cities;
* improved income opportunities through wet- or dry-season agriculture, capture fishing or aquaculture; or
* both simultaneously, although costs and benefits may accrue to different groups.

A recent example of a flood disaster is that of the Chao Phrya Basin, central Thailand in 2011, caused by a combination of bad decisions on reservoir management, copious late-season precipitation, and the inadequacy of the Bangkok flood-control system (Komori et al., 2012). There is danger of similar, costly, man-made floods along a cascade of dams in the Mekong Basin if the dams are full, late-season rainfall is high, and the dam operators do not communicate and coordinate water release from the dams (MK3) (Ward et al., 2012).

We found in the Ganges Basin Focal Project that,

> Floods are a common feature . . . Flooding in rivers is mainly caused by inadequate capacity within the banks of the rivers to contain higher flows [that may be generated by exceptional rainfall or exceptional runoff resulting from land-use imposed changes in soil structure and thus water infiltration], riverbanks erosion and silting of riverbeds, landslides leading to obstruction of flow and change in the river course, poor natural drainage due to flat floodplains and occurrence of coastal cyclones, and intense rainfall events . . . Among the South Asian countries, India is more vulnerable to flood events, followed by Bangladesh.
> (PN60) (Mishra, 1997; Sharma, 2010)

The Limpopo Basin suffers severe floods, interspersed with droughts. Although the basin is on average water scarce, there are peak-rainfall periods during which large amounts of runoff flow from the basin quickly as floods. The flood flows are not captured and are not available to agriculture (PN62) (Sullivan and Sibanda, 2012).

Not all floods are harmful. Smallholder communities use seasonally flooded lands in Bangladesh for aquaculture to generate substantial income (PN35) (Sheriff, 2010). The wet-season flood and dry-season ebb of the Mekong provides a productive capture fishery used by smallholders in the Tonle Sap in Cambodia (PN58, MK5 and MK2) (Kirby et al., 2010b; Mainuddin et al., 2011; Kura, 2012; Pukinskis and Geheb, 2012). The lower Mekong Basin yields about 4.5 mt/yr of fish and aquatic products worth US$3.9–7 billion/yr, with fisheries contributing to the diversification of livelihoods of the poor. The annual flood–ebb pulse opens up new feeding areas for fish to feed and triggers migration in some fish species (Pukinskis and Geheb, 2012).

The Yellow River and the Niger have similar seasonally flooded fisheries (PN69) (Kam, 2010; Béné et al., 2009). Seasonal flooding of the Nile was important to cropping in Egypt, especially in the delta, but no longer occurs downstream of the Aswan high dam.

Access

Our definition of farm-level water scarcity includes who has access to the water resources. Because of conflicting interests among water users, it can be difficult to define who the right people are. Conflict over access to water can occur at the community, landscape, catchment, basin and regional levels, or even internationally. Here we only give a few examples as Chapter 6 on the contributions of research to understanding and strengthening institutions for equitable water resource management discusses water access at greater length.

In the Limpopo Basin, access to water resources is inequitable with larger commercial farmers having preference over smallholders (PN62) (Alemaw et al., 2010). In the Mekong Basin, conflicts in the use of water to generate hydropower to the detriment of agriculture and fisheries have been researched in both CPWF phases (PN67, MK5, MK4, MK3 and others) (Dore et al., 2010; Joffre et al, 2011; Pukinskis and Geheb, 2012; Sajor, 2012). The challenge has been to find ways to protect farming, fisheries and ecosystem services even as planning, construction and operation of hydropower dams go ahead (Ziv et al., 2012).

Improvements in productivity can sometimes intrude on access to water. Improved community-managed aquaculture during the wet season in season-ally flooded areas in Bangladesh precedes a dry-season crop. As the economic success of aquaculture became apparent, private investors began to compete to lease the fishing rights, threatening community access (PN35) (Sheriff et al., 2010; Ratner et al., 2012).

Water access is linked to seasonality and water quality, often in complex ways. In coastal Vietnam, some farmers wanted freshwater to grow rice while others wanted saline water to grow shrimp at different times during the year and at different places. Researchers analyzed land- and water-use options using modeling. They defined suitable areas both for rice and for shrimp, which effectively resolved conflict and fostered intensification and diversification of the farming system (PN10) (Tuong and Hoanh, 2009).

Self-supply and informal arrangements flexible enough to cope with the harsh climate governed traditional access to rural water in South Africa (and elsewhere). When legislation established formal water rights, the "reform basically [dispossessed poor rural communities] from their current and future claims to water" (PN66) (van Koppen, 2010).

Water access often has an upstream–downstream dimension. For example, the proliferation of upstream small reservoirs in the Volta might threaten flows into the Akosombo dam and the hydropower it generates. Project PN46 found that "the collective downstream impact of the present number of small reservoirs is minimal"; that "[even] after quadrupling the present number of small reservoirs, their combined impact will be less than 1% of the total water balance." It concluded that the "reservoirs do not deprive downstream users of the water for hydropower, agriculture, and environmental flows" (Liebe, 2002; Andreini et al., 2010).

Similarly, in Ecuador, Quito's water company planned to increase with-drawals from the Quijos River to meet increased urban demand. This raised concerns about lessened downstream flow and its consequences on economic activity. Project Andes 2 showed that the middle part of the watershed receives enough rainfall to replace most of the upstream withdrawals so that down-stream activities would be little affected. Stakeholders will use this information to negotiate appropriate levels of compensation (Quintero et al., 2012).

Finally, there has been a long-standing debate between upstream and downstream countries in the Nile, over the effects downstream of upstream development of hydropower and large-scale irrigation. A recent book based on a CPWF project concluded that "there is enough water to supply dams and irrigate parched agriculture in all ten [Nile Basin] countries—but policymakers risk turning the poor into water 'have-nots' if they do not enact inclusive water management policies" (Awulachew et al., 2012a).

WP revisited

The CPWF proposal in 2001 defined low WP as an important problem (see Chapter 1 for an overview of the creation of the CPWF and the activities of the international water community). Indeed the objectives of many projects approved in Phase 1 of the CPWF had as their primary objective to raise WP of systems and sought to understand the reasons why WP was so low. Here we revisit the concept of WP by examining how well the emphasis on it allowed the CPWF to address its main objectives, that is, what did we learn about using WP as an important indicator performance?

The first question is why WP and not some other measure such as land, labor, capital or total factor productivity? While authors had noticed that WP was not necessarily a factor that farmers could easily accept (Luquet et al., 2005), at the time the CPWF was conceived, it was in response to a widely held view that a global water crisis was looming. This was supported by the address of the UN Secretary-General to the General Assembly in 2001 calling for "more crop per drop." Global population was forecast to increase by 50 percent by 2050 from the 6 billion reached in October 1999, and it was thought that there would not be enough water to grow the food that would be needed. It was therefore reasonable to focus on WP as one way to address the crisis, and incorporate it as a main objective of the CPWF. The CPWF proposal emphasized the importance of WP (CPWF Consortium, 2002), which was still regarded as being of great importance for the rest of the decade (Rijsberman, 2004; Molden, 2007).

Initial concept

"Productivity of water is related to the value or benefit derived from the use of water" (Molden, 1997; Molden et al., 2003). Starting from the point of view of irrigation, water is classified on its utility, as to whether it is depleted (removed from the system as by crop ET, flows to a sink, or becomes so polluted as to be unusable), or whether it is outflow, which may or may not be committed to some downstream use. The basin WP estimate will include the WP of the downstream use if the outflow is used consumptively. In the context of irrigation, WP is straightforward with the denominator as depleted water and the numerator being either the yield of the crop, the saleable value of that yield, or some other relevant measure such as energy content (yield of calories).

Authors have broadened the WP concept to include rainfed agriculture, grazing animals and aquaculture. Each of these presents difficulties in deciding what to use as the denominator. For example, of the rain that falls on a crop, some evaporates *in situ*, a fraction enters the soil and a variable part is runoff, which is likely to become blue water. The fraction that enters the soil is either taken up by plants and transpired, or percolates to depth where it may replenish an aquifer, which may contribute to the blue water of stream flow. Given these possibilities, what fraction of the precipitation should we use as the denominator in calculating WP? The answer depends on the scale of comparison, often using annual or seasonal rainfall or some estimate of ET and ignoring runoff and downstream use. Because the conditions in different basins are rarely the same, it is usually not valid to compare WP between basins, although it can be used with caution to indicate relative efficiency of water use. The same arguments apply to intra-basin comparisons.

Different measures of WP

All terrestrial water originates from precipitation; even that stored in aquifers came from historic precipitation. Hydrologists are principally concerned with the utilization of blue water in managed irrigation systems. In rainfed systems, the denominator should be ET, but this is difficult to estimate even at the level of an experimental plot. It becomes more problematic as the scale increases to the field, or the farm, but at the broader landscape or basin level, it can be estimated by remote sensing (Vidal and Perrier, 1990; Ahmad et al, 2008). At the level of a basin, authors typically use precipitation, either annual, or for the growing season where there is more than one crop a year. In these cases, WP needs to be interpreted with caution (Ogilvie et al., 2012).

Where water does not limit crop production, as in southern Nigeria, for example, factors other than water, such as soil fertility, control WP. The paradox is that apparent WP is higher in rainfed systems with lower precipitation than it is in areas where rainfall is adequate or more abundant. The paradox arises because the divisor is received precipitation. Crops in wetter areas in general use a lower proportion of the rainfall than those in drier areas. Moreover, if farmers use risk-avoidance strategies, such as low sowing densities, that give some yield in bad years but cannot give high yields in good years, average long-term WP will be low (Terrasson et al., 2009). For this reason WP of crops in rainfed systems, especially those where only a small fraction of rainfall goes through agriculture, also needs to be interpreted with caution.

WP based on crop yield in irrigated systems is straightforward, although we need to be careful when we consider higher-value crops, where the higher value may not offset lower yields. Then there is the converse example where low-yielding crops, such as cotton in the Gezira, are grown upstream, potentially limiting the water available for more valuable, higher-yielding crops downstream in the Nile valley and its delta.

Water productivity of herbivores is difficult because they consume only a small fraction of the available herbage (Peden et al., 2009). In well-managed tropical pastures, utilization by cattle is rarely more than 30 percent. It is at least tenfold less in extensively grazed rangeland. In estimating WP in aquaculture, losses to evaporation and infiltration of the ponds are the denominator, unless the outflow is too contaminated to use downstream. WP in capture fisheries is debatable, because water lost to evaporation or infiltration would be lost anyway, but if preservation of the fishery resource prevents another use such as hydropower, there is a cost in foregone development.

Utility of WP

As the CPWF progressed, and with more analysis, the objective remained to improve WP, "more crop per drop," but the limitations of the approach as an end in itself became clearer. It is relatively straightforward to measure crop WP with a combination of satellite data "to estimate both crop production and consumptive water use" (Cai et al., 2012). It is more difficult to estimate WP of livestock systems and capture fisheries, both of which need "development of concepts and methodology" (Cai et al., 2012).

Limitations of WP

Because of the complexity in measuring it and interpreting the data, we conclude that WP has limited usefulness as an objective. Nevertheless, WP is a useful diagnostic tool, which with other data can identify bright spots of high productivity and hot spots of low productivity per unit of water depleted in irrigation systems. It is less useful to identify inefficient rainfed cropping systems, except in the broadest sense; for example, rainfed agriculture in West Africa clearly performs poorly. It is more useful to analyze the reasons why it performs so poorly.

WP still remains an important identifier of efficiency in irrigated systems; for instance the WP of the huge Gezira irrigation area in Sudan was low as a result of central control, which prescribed crop management and required that the tenant farmers plant 20 percent of their land to cotton. Recent administrative changes allow some crop diversification and WP is improving, but remains low (Awulachew et al., 2012a; 2012b). In contrast, there are bright spots in the Indian Punjab in the Ganges Basin, where WP is close to its practical maximum (Sharma et al., 2012). When we apply the concept of WP to rainfed systems, the results are trivial because only a small proportion of the precipitation is used by agriculture.

There are wide variations in WP in rainfed systems within basins, for example, the Volta (Lemoalle and de Condappa, 2012), the Karkheh (Ahmad and Giordano, 2012) and the Limpopo (Sullivan and Sibanda, 2012). Each basin needs careful analysis to identify the causal factors, which differ between basins, so that it is impossible to make blanket recommendations. Focus on WP can (and did) overshadow other, equally important indicators of productivity and livelihoods, which partly explains why the CPWF focus broadened from water scarcity to include development challenges.

Final observations

Authors often write that "drought tolerance" can improve WP, although the term is rarely explained. Certainly crops or crop varieties that are able to survive short droughts without too much damage are likely to give yields that are more reliable in droughty environments than those that cannot. However, gains in WP through new germplasm are most likely when plants are capable of yielding well under favorable climatic conditions as well as being tolerant of drought and other abiotic stresses. Reliable yields under dry conditions are only half of the story.

Much of the improvements in yield (and hence WP) last century were achieved by plant breeders who changed harvest index, that is, the proportion of the commercial product (often grain) in the total yield (Gifford and Evans, 1981; Bennett, 2003). They achieved this by breeding short-strawed rice and wheat, and hybrid sorghum and maize with shorter stature and reduced root systems, which were possible on good soils with precision fertilizer placement. In rainfed agriculture in many of the CPWF's target basins, smallholder farmers typically grow rustic varieties with low harvest indices. High-yielding plant varieties and fertilizer can increase WP, but the institutional and sociological problems that constrain farmers from adopting them are the key issues. There is also the issue of varietal adaptation: modern varieties do sometimes perform poorly when grown under stresses to which they lack adaptation. Nonetheless, as a general rule increased WP tends to accompany increased land productivity, and both require plant types that can yield well under favorable conditions as well as tolerating unfavorable conditions.

As understanding accumulated during the currency of the CPWF, it recognized that factors other than WP itself were more important to livelihoods and food security, especially of the poor. It concluded that WP was a useful indicator for some purposes, but was not the critically important factor that it was assumed to be when the CPWF was initiated. The CPWF broadened its agenda to focus on development challenges in basins related to water. It came to see that addressing water scarcity was a means of helping achieve broader development goals, including reducing poverty, rather than an end in itself. The change is an example of how learning helped the CPWF to grow and evolve, adjusting its priorities and research questions as its understanding of the issues improved.

Water, poverty and water poverty

Kemp-Benedict et al. (2012) summarized the variables used to estimate poverty in the BFP's ten basins. Water scarcity was not strongly correlated with poverty, which highlights the danger of assuming that there is a simple association between water availability and poverty. Other variables that do explain variations in poverty are those responsible for basic livelihood support, including access to water, protection from hazards such as drought and flood, and the ability to produce increased amounts of high-quality food.

Evidence suggests that poverty is more dependent on the stage of development of the basin's economy (Cook et al., 2012). At their least-developed phase, populations in basins are low in proportion to the resources available. In this case, poverty is more strongly related to the absence of basic services such as safe water, sanitation, health care, education, finance, markets or farming inputs. Pressure on resources increases as a basin's economy develops during the transitional phase, so that both scarcity of and access to water become important. As economies move toward industrialization, these deficiencies are corrected, but some sectors of the populations are left behind in relative poverty, showing that the benefits from growth do not trickle down to the whole population, especially to the most vulnerable. This pattern of economic evolution parallels a general movement from informal to formal governance structures. This makes formal policy interventions less effective for less-developed basins. Similarly, as incomes increase there are more livelihood opportunities, which blurs the relationship between water and poverty. In a nutshell, when economies develop, we see a weakening of the link between the provision of natural resources and livelihood outcomes.

Socio-economic development changes the manner in which food and water systems utilize ecosystem services (freshwater, soil formation, nutrient cycling) within geographically diverse river basins. We order these with respect to development along a single trajectory (Byerlee et al., 2009). This trajectory is

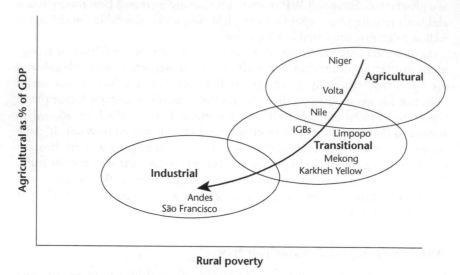

Figure 2.2 CPWF basins ordered according to rural poverty and agricultural contribution to gross domestic product (GDP).

Source: Cook et al., 2012.

defined by the level of rural poverty or of urbanization and is strongly related
to the contribution of agriculture to GDP growth (Figure 2.2). This concept
classifies economies as they move from conditions that are described as
agricultural, through transitional to industrial. Here we organize observations
from ten river basins to focus on the characteristics of developing food and
water systems, and of the ecosystem services that support them.

Agricultural economies

Agricultural economies are characterized by a high dependence on agriculture
for GDP and widespread rural poverty. These conditions predominate in the
African basins (Niger, Volta and Nile, except Egypt) but also occur in parts of
other basins, such as in upper parts of the Mekong and Yellow rivers. Many of
these basins include semi-arid areas, but analyses by Awulachew et al. (2010),
Lemoalle and de Condappa (2012) and Ogilvie et al. (2012) indicate that the
relationship of poverty with water availability is weak. Other factors are more
important, including the vulnerability to drought and flood, and lack of access
to water and other benefits such as roads, safe water and sanitation.

Despite the influence of drought, water resources are hardly developed:
irrigation consumes less than 1 percent of water resources and covers less than
1 percent of the landscape in the Niger and the Volta Basins (Lemoalle and de
Condappa, 2012; Ogilvie et al., 2012). Even in the Nile, irrigation accounts
for less than 4 percent of the water balance, virtually all restricted to Egypt and

the Gezira in Sudan (Awulachew et al., 2010; Kirby et al., 2010c). Rainfed agriculture dominates and is generally of very low productivity. Rural liveli-hoods depend on a diversity of low-intensity activities that reduce risk. Poverty is widespread but absolute numbers are low because of low population densities. Birth and mortality rates are very high. Poverty is associated with lack of access to resources and vulnerability to hazards of drought, flood, malaria, and other—often water-borne—diseases.

Transitional economies

Transitional economies are identified by a reduced dependence on agriculture for GDP growth and a coincident reduction in levels of rural poverty, even though in some basins, such as the Ganges and Yellow, absolute numbers below the poverty line remain very large, for example, more than 220 million in the Ganges (Sharma et al., 2010). In these basins, vast numbers of farmers are supported by irrigation. In some areas (e.g. the Indian Punjab or Shandong province in the Yellow River) this is extremely productive and has a clear impact in reducing rural poverty and on national food security and economic activity. The Yellow River basin produces 14 percent of China's grain and about 14 percent of GDP while consuming only about 2 percent of the nation's water (Ringler et al., 2012). Irrigation has a clear impact on poverty alleviation, and reduces major sources of vulnerability.

Provision of basic necessities that accompany the development of agricultural systems reduces mortality, and most transitional economies show a substantial decline in fertility. As economies develop further, the competition for water resources for urban and industrial supply intensifies. In the Ganges, Indus, Karkheh and Yellow rivers, as well as the Nile delta, this is a cause of major tension, especially if irrigated agriculture is locked into relatively low-value production of commodities. Rainfed agricultural productivity also increases in response to demand, but generally value-adding remains low. During this phase, regulating ecosystem services suffer widespread loss since institutions aimed at preserving ecosystem resilience are rudimentary or powerless, while those supporting resource exploitation are very powerful.

By the end of this phase, aquatic ecosystems will be substantially modified, as seen in the Mekong, Ganges, Niger, São Francisco and Karkheh Basins. In the Yellow River, capture fisheries have been eliminated. Elsewhere they are likely to be severely reduced or replaced by aquaculture. Extensive livestock systems will have been replaced in part by more-intensive production. In summary, this phase is characterized by a major expansion in productivity but also widespread reduction in the range of ecosystem services.

The industrial classification

The industrial classification applies to basins in which agriculture contributes 5 percent on average to GDP growth and poverty is mostly urban (World

Bank, 2007). These basins are characterized by growth of value and a broadening of institutions to support more resilient growth. These conditions occur within parts of the São Francisco and Andes, as well as regionally in the Mekong, Limpopo and Yellow River basins (Kirby et al., 2010b; Mulligan et al., 2012b; Ringler et al., 2012; Sullivan and Sibanda, 2012). Now value-adding of non-agricultural activities dominates national economies but agriculture remains an important contributor to rural livelihoods. Declining fertility stabilizes population, although total demand on ecosystem services continues to grow as per capita consumption of food and energy increases, as does the demand for meat and dairy products.

Poverty persists in pockets, amongst groups who are excluded from urban-centered prosperity or amongst the urban poor, many of whom may be recent migrants from agriculture. Availability of adequate, clean water may be recognized as a major constraint to long-term development. Political discourse reflects a growing recognition of the reliance on regulating ecosystem services and institutions to distribute risks and benefits more widely. Valuation and trading of water and other ecosystem services emerges, although by this stage, degradation of some ecosystem services may have proceeded to a point at which loss is permanent.

Conclusions

This analysis from ten river basins at different stages of development indicated that stresses are emerging in some basins as a result of unsustainable exploitation of natural resources, which are affecting ecosystem services. In contrast, we found that underutilized productive capacity, unequal access and unbalanced development are more widespread. While the underutilized capacity is sufficient—in theory—to satisfy the food and energy demands of populations to 2050, the unequal access and unbalanced development yield, in turn, relative poverty and exclusion in all basins. We therefore consider three conditions as necessary to achieve better income and food security. First, we see a need for a major increase in agricultural productivity, particularly in rainfed systems. Second, there needs to be greater sharing of benefits and risks between different groups of people to capitalize on the collective benefits. Third, changes toward more collaborative use of ecosystem services require long-term investment, which must be underwritten by political discourse. The discourse must recognize the development process and how development impacts the ecosystem services that are used to support it.

Having observed that poverty is more dependent on the stage of development of the country's or basin's economy, the CPWF reoriented its approach toward addressing specific development issues or *challenges* in basins, which has led to designing the approaches described in further chapters.

We know that our conclusions are based on data that are often less detailed and less reliable than we would like, even though they are the best available, but we did not attempt to draw detailed conclusions. The conclusions drawn are general enough that lack of detailed data is not a problem. In the future, better data may be available from remote sensing, such as the Tropical Rainfall Measuring Mission. Socio-economic data will always depend on data from the countries concerned.

Acknowledgments

We acknowledge helpful comments from Tilahun Amede, Kim Geheb, Nancy Johnson, Eric Kemp-Benedict, Ilse Pukinskis, Marcela Quintero and two anonymous reviewers.

Notes

1   "Blue water" is the precipitation that enters lakes, rivers and groundwater. It is the main source of water used for industry, domestic purposes and irrigation. Only 30–35 percent of precipitation becomes blue water. "Green water" is the precipitation that enters the soil, is taken up by plants and transpired back to the atmosphere. Rainfed agriculture depends on green water, which is about 65 percent of all precipitation. For further explanation, see wmc.landfood.ubc.ca/webapp/VWM/course/global-water-challenges/green-and-blue-water-cycle (accessed 9 April 2014).
2   There were the nine CPWF benchmark basins, Andes, Indus-Ganges, Karkheh, Limpopo, Mekong, Nile, São Francisco, Volta and Yellow.
3   The ten were the nine CPWF benchmark basins plus the Niger.
4   At times we refer to specific projects. In this case, PN10 means "Project Number 10" from Phase 1 (see the Appendix for a complete list of projects).
5   Arsenic contamination of groundwater is a serious problem in many areas of Bangladesh. Groundwater use is uncommon in the polder study areas and groundwater irrigation is not part of the set of innovations introduced because of salt intrusion in groundwater as well as the possibility of arsenic contamination.

References

Abaidoo, R. C., Keraita, B., Amoah, P., Drechsel, P., Bakang, J., Kranjac-Berisavljevic, G., Konradsen, F., Agyekum, W. and Klutse, A. (2009) *Safeguarding public health concerns, livelihoods and productivity in wastewater irrigated urban and periurban vegetable farming*, CPWF Project Report Series PN38, CGIAR Challenge Program on Water and Food, Colombo, Sri Lanka.

Ahmad, M. D., Turral, H. and Nazeer, A. (2008) 'Diagnosing irrigation performance and water productivity through satellite remote sensing and secondary data in a large irrigation system of Pakistan', *Agricultural Water Management*, doi: 10.1016/j.agwat.2008.09.017.

Ahmad, M. and Giordano, M. (2012) 'The Karkheh River basin: The food basket of Iran under pressure', in M. Fisher and S. Cook (eds) *Water, food and poverty in river basins: Defining the limits*, Routledge, London, pp. 59–81.

Alemaw, B. F., Scott, K. and Sullivan, A. (2010) *Water availability and access in the Limpopo Basin*, Working Paper: Basin Focal Project Series, BFP-L01, CGIAR Challenge Program on Water and Food, Colombo, Sri Lanka.

Amede, T., Kassa, H., Zeleke, G., Shiferaw, A., Kismu, S., and Teshome, M. (2007) 'Working with communities and building local institutions for sustainable land management in the Ethiopian Highlands', *Mountain Research and Development* vol. 27, no. 1, pp. 15–19.

Andreini, M., Schuetz, T., Senzanje, A., Rodriguez, I., Andah, W., Cecchi, P., Boelee, E., van de Giesen, N., Kemp-Benedict, E. and Liebe, J. (2010) *Small multi-purpose reservoir ensemble planning*, CPWF Project Report Series PN46, CGIAR Challenge Program on Water and Food, Colombo, Sri Lanka.

Awulachew, S., Ahmed, A., Haileselassie, A., Yilma, A., Bashar, K., McCartney, M. and Steenhuis, T. (2010) *Improved water and land management in the Ethiopian highlands and its impact on downstream stakeholders dependent on the Blue Nile*, CPWF Project Report Series PN19, CGIAR Challenge Program on Water and Food, Colombo, Sri Lanka.

Awulachew, S., Rebelo, L.-M. and Molden, D. (2012a) 'The Nile Basin: Tapping the unmet agricultural potential of Nile waters', in M. Fisher and S. Cook (eds) *Water, food and poverty in river basins: Defining the limits*, Routledge, London, pp. 160–191.

Awulachew, S., Smakhtin, V., Molden, D. and Peden D. (2012b) *The Nile River basin: Water, agriculture, governance and livelihoods*, Routledge, London.

Béné, C., Kodio, A., Lemoalle, J., Mills, D., Morand, P., Ovie, S., Sinaba, F. and Tafida, A. (2009) *Participatory diagnosis and adaptive management of small-scale fisheries in the Niger River basin*, CPWF Project Report Series PN72, CGIAR Challenge Program on Water and Food Colombo, Sri Lanka.

Bennett, J. (2003) 'Opportunities for increasing water productivity of CGIAR crops through plant breeding and molecular biology' in J. W. Kijne, R. Barker and D. Molden (eds) *Water productivity in agriculture: Limits and opportunities for improvement*, CAB International, Wallingford, UK, pp. 103–126.

Block, P. (2008) *Mitigating the effects of hydrologic variability in Ethiopia: An assessment of investments in agricultural and transportation infrastructure, energy and hydroclimatic forecasting*, CPWF Research for Development (R4D) Series 01, CGIAR Challenge Program on Water and Food Colombo, Sri Lanka.

Boelee, E., Cecchi, P., Kone, A., Cecchi, P. and Koné, A. (2009) *Health impacts of small reservoirs in Burkina Faso*, IWMI Working Paper 136, International Water Management Institute, Colombo, Sri Lanka.

Bouman, B. A. M. (2008) *Developing a system of temperate and tropical aerobic rice in Asia (STAR)*, CPWF Project Report Series PN16, CGIAR Challenge Program on Water and Food, Colombo, Sri Lanka.

Byerlee, D., de Janvry, A. and Sadoulet, E. (2009) 'Agriculture for development: Toward a new paradigm', *Annual Review of Resource Economics* vol. 1, pp. 15–31.

Cai, X., Molden, D., Mainuddin, M., Sharma, B., Ahmad, M. and Karimi, P. (2012) 'Producing more food with less water in a changing world: Assessment of water productivity in 10 major river basins', in M. Fisher and S. Cook (eds) *Water, food and poverty in river basins: Defining the limits*, Routledge, London, pp. 280–300.

Candelo, C., Cantillo, L., Gonzalez, J., Roldan, A. and Johnson, N. (2008) 'Empowering communities to co-manage natural resources: Impacts of the *Conversatorio de Acción Ciudadana*', in *Proceedings of the CGIAR Challenge Program on Water and Food 2nd International Forum on Water and Food, Addis Ababa, Ethiopia, 10–14 November, 2008*, CGIAR Challenge Program on Water and Food, Colombo, Sri Lanka.

Castillo, E. G., Tuong, T. P., Ismail, A. M. and Inubushi, K. (2007) 'Response to salinity in rice: Comparative effects of osmotic and ionic stresses', *Plant Production Science*, vol. 10, no. 2, pp. 159–170.

Cecchi, P. and Sanogo, S. (2012) 'CPWF-V3: Ecological externalities', in *Volta BDC Science Workshop, Ouagadougou, 3–5 July 2012*, CGIAR Challenge Program on Water and Food, Colombo, Sri Lanka.

Chartres, C., and Varma, S. (2011) *Out of water: From abundance to scarcity and how to solve the world's water problems*, FT Press Upper Saddle River, NJ.

CIMMYT (1999) 'Risk management for maize farmers in drought-prone areas of Southern Africa', in *Proceedings of a Workshop on Maize in Drought-Prone Areas, Kadoma Ranch, Zimbabwe, 1–3 October, 1997*, Centro Internacional de Mejoramiento de Maíz y Trigo, Texcoco CP, Mexico, International Crops Research Institute for the Semi-Arid Tropics, Hyderabad, India and Danish International Development Agency, Copenhagen, Denmark.

Clanet, J. C. and Ogilvie, A. (2009) *Basin Focal Project Niger*, CPWF Project Report PN64. CGIAR Challenge Program on Water and Food, Colombo, Sri Lanka.

Cook, S., Fisher, M., Tiemann, T. and Vidal, A. (2012) 'Water, food and poverty: Global- and basin-scale analysis', in M. Fisher and S. Cook (eds) *Water, food and poverty in river basins: Defining the limits*, Routledge, London, pp. 239–254.

Cooper, P. J. M., Dimes, J., Rao, K. P. C., Shapiro, B., Shiferaw, B. and Twomlow, S. (2008) 'Coping better with current climatic variability in the rain-fed farming systems of sub-Saharan Africa: An essential first step in adapting to future climate change?', *Agriculture, Ecosystems and Environment*, vol. 126, no. 1, pp. 24–35.

CPWF (2012) *Addressing public health issues in urban vegetable farming in Ghana*, Outcome Stories Series. CGIAR Challenge Program on Water and Food, Colombo, Sri Lanka.

CPWF Consortium (2002) *CGIAR Challenge Program on Water and Food—Full proposal*, Colombo, Sri Lanka.

Dimes, J., Twomlow, S., Rusike, J., Gerard, B., Tabo, R., Freeman, A. and Keatinge, J. D. H. (2005) 'Increasing research impacts through low-cost soil fertility management options for Africa's drought-prone areas', in *Proceedings of the International Symposium for Sustainable Dryland Agriculture Systems, 2–5 December 2003 Niamey, Niger*, International Crops Research Institute for the Semi-Arid Tropics, Niamey, Niger.

Dore, J., Molle, F., Lebel, L., Foran, T. and Lazarus, K. (2010) *Improving Mekong water resources investment and allocation choices*, CPWF Project Report Series PN67, CGIAR Challenge Program on Water and Food, Colombo, Sri Lanka.

Eastham, J., Kirby, M., Mainuddin, M. and Thomas, M. (2010) *Water use accounts in CPWF basins: Simple water use accounting of the Limpopo Basin*. CPWF Working Paper Basin Focal Project Series 06, CGIAR Challenge Program on Water and Food, Colombo, Sri Lanka.

Estrada, R. D., Quintero, M., Moreno, A. and Ranvborg, H. M. (2009) *Payment for environmental services as a mechanism for promoting rural development in the upper watersheds of the tropics*, CPWF Project Report Series PN22, CGIAR Challenge Program on Water and Food, Colombo, Sri Lanka.

Falkenmark, M. (1997) 'Society's interaction with the water cycle: A conceptual framework for a more holistic approach', *Hydrological Sciences*, vol. 42, pp. 451–466.

Falkenmark, M. and Rockström, J. (2010) 'Building water resilience in the face of global change: From a blue-only to a green–blue water approach to land-water management', *Journal of Water Resources Planning and Management*, vol. 136, no. 6, pp. 606–610.

Fisher, M. J. and Cook, S. E. (eds) (2012) *Water, food and poverty in river basins: Defining the limits*, Routledge, London, 400pp.

FUNDESOT (2012) *Montaje de parcelas con los sistemas de agricultura de conservación en Carmen de Carupa (Colombia)*, FUNDESOT, Cota, Colombia.

Gallopín, G. and Rijsberman, F. (2000) 'Three global water scenarios', *International Journal of Water* vol. 1, no. 1, pp. 16–40.

Gifford, R. M. and Evans, L. T. (1981) 'Photosynthesis, carbon partitioning, and yield', *Annual Review of Plant Physiology*, vol. 32, pp. 485–509.

Grey, D. and Sadoff, C. (2002) 'Water resources and poverty in Africa: Breaking the vicious circle', Paper presented at the inaugural meeting of the African Ministers' Council on Water (AMCOW), at Abuja, Nigeria 30 April, 2002, africanwater.org/Documents/amcow_wb_speech.pdf (accessed 3 June 2013).

Harrington, L. and Tow, P. (2011) 'Types of rainfed farming systems around the world', in P. Tow, I. Cooper, I. Partridge and C. Birch (eds) *Rainfed Farming Systems*, Springer, Dordrecht, Netherlands, pp. 45–74.

Humphreys, E. (2012) *Project Ganges 2: Productive, profitable and resilient agriculture and aquaculture systems*, CPWF Six-Monthly Project Reports for April 2012–September 2012, CGIAR Challenge Program on Water and Food, Colombo, Sri Lanka.

Islam, M. R., Salam, M. A., Bhuiyan, M. A. R., Rahman, M. A. and Gregorio, G. B. (2008) 'Participatory variety selection for salt tolerant rice', *International Journal of BioResearch* vol. 4, no. 3, pp. 21–25.

Ismail, A. M. (2009) *Development of technologies to harness the productivity potential of salt-affected areas of the Indo-Gangetic, Mekong, and Nile River basins*, CPWF Project Report Series PN07, CGIAR Challenge Program on Water and Food, Colombo, Sri Lanka.

Joffre, O., Rasaenan, T., Meynell, P. J., Baran, E., Ketelsen, T., Paradissometh and Hong, T. (2011) 'Scoping for sustainable multiple use of water in cascades of hydropower schemes', in *Proceedings of the CGIAR Challenge Program on Water and Food 3rd International Forum on Water and Food, Tshwane, South Africa, 14–17 November 2011*, CGIAR Challenge Program on Water and Food, Colombo, Sri Lanka.

Johnson, N., Garcia, J., Rubiano, J. E., Quintero, M., Estrada, R. D., Mwangi, E., Morena, A., Peralta, A. and Granados, S. (2009) 'Water and poverty in two Colombian watersheds', *Water Alternatives*, vol. 2, no. 1, pp. 34–52.

Kam, S. P. (2010) *Valuing the role of living aquatic resources to rural livelihoods in multiple-use, seasonally inundated wetlands in the Yellow River basin of China for improved governance*, CPWF Project Report Series PN69, CGIAR Challenge Program on Water and Food, Colombo, Sri Lanka.

Kemp-Benedict, E., Cook, S., Allen, S. L., Vosti, S., Lemoalle, J., Giordano, M., Ward J. and Kaczan, D. (2012) 'Connections between poverty, water, and agriculture: Evidence from ten river basins', in M. Fisher and S. Cook (eds) *Water, food and poverty in river basins: Defining the limits*, Routledge, London, pp. 363–378.

Kirby, M., Mainuddin, M. and Eastham, J. (2010a) *Water-use accounts in CPWF basins: Model concepts and description*, CPWF Working Papers, Basin Focal Projects Series BFP01, CGIAR Challenge Program on Water and Food, Colombo, Sri Lanka.

Kirby, M., Mainuddin, M. and Eastham, J. (2010b) *Water-use accounts in CPWF basins: Simple water-use accounting of the Mekong Basin*, CPWF Working Papers, Basin Focal Projects Series BFP02, CGIAR Challenge Program on Water and Food, Colombo, Sri Lanka.

Kirby, M., Mainuddin, M. and Eastham, J. (2010c) *Water-use accounts in CPWF basins: Simple water-use accounting of the Nile Basin*, CPWF Working Papers, Basin Focal Projects Series BFP03, CGIAR Challenge Program on Water and Food, Colombo, Sri Lanka.

Komori, D., Nakamura, S., Kiguchi, M., Nishijima, A., Yamazaki, D., Suzuki, S., Kawasaki, A., Oki, K. and Oki, T. (2012) 'Characteristics of the 2011 Chao Phraya River flood in Central Thailand', *Hydrological Research Letters*, vol. 6, pp. 41–46, doi: 10.3178/HRL.6.41.

KPMG (2012) *Sustainable insight—Water scarcity: A dive into global reporting trends*. KPMG International Cooperative, Swiss based.

Kura, Y. (2012) *Project Mekong 2: Water valuation*, CPWF Six-Monthly Project Reports for April 2012–September 2012, CGIAR Challenge Program on Water and Food, Colombo, Sri Lanka.

Lemoalle, J. (2008) *Basin Focal Project Volta*, CPWF Project Report Series PN55, CGIAR Challenge Program on Water and Food, Colombo, Sri Lanka.

Lemoalle, J. and de Condappa, D. (2012) 'Farming systems and food production in the Volta Basin', in M. Fisher and S. Cook (eds) *Water, food and poverty in river basins: Defining the limits*, Routledge, London, pp. 192–217.

Liebe, J. (2002) 'Estimation of water storage capacity and evaporation losses of small reservoirs in the Upper East Region of Ghana', Diploma thesis, University of Bonn, Germany.

Luquet, D., Vidal, A., Smith, M. and Dauzat, J. (2005) '"More crop per drop": how to make it acceptable for farmers?', *Agricultural Water Management*, vol. 76, no. 2, pp. 108–119, doi:10.1016/j.agwat.2005.01.011.

Mainuddin, M., Kirby, M. and Chen, Y. (2011) *Fishery productivity and its contribution to overall agricultural production in the Lower Mekong River Basin*, CPWF Research for Development (R4D) Series 03. CGIAR Challenge Program on Water and Food, Colombo, Sri Lanka.

Mee, L. and Adeel, Z. (2012) *Science-policy bridges over troubled waters—Making science deliver greater impacts in shared water systems*. United Nations University Institute for Water, Environment and Health (UNU-INWEH), Hamilton, Canada.

Middleton, N. and Thomas, D. (1997) *World atlas of desertification*, Arnold, Hodder Headline, London.

Mishra, D. K. (1997) 'The Bihar flood story', *Economic and Political Weekly*, vol. 32, no. 35, pp. 2206–2217.

Molden, D. (1997) *Accounting for water use and productivity. System-wide Initiative on Water Management Paper 1*, International Irrigation Management Institute, Colombo, Sri Lanka.

Molden, D., Murray-Rust, H., Sakthivadivel, R. and Makin, I. (2003) 'A water-productivity framework for understanding and action', in J. W. Kijne, R. Barker and D. Molden (eds) *Water productivity in agriculture: Limits and opportunities for improvement*, CAB International, Wallingford, UK, pp. 1–18.

Molden, D. (ed) (2007) *Water for food, water for life: A comprehensive assessment of water management in agriculture*, Earthscan, London and International Water Management Institute, Colombo, Sri Lanka.

Mukherji, A. (2012) *Project Ganges 3: Water governance and community-based management*. CPWF Six-Monthly Project Reports for April 2012–September 2012, CGIAR Challenge Program on Water and Food, Colombo, Sri Lanka.

Mulligan, M., Rubiano, J., Hyman, G., Leon, J. G., Saravia, M., White, D., Vargas, V., Selvaraj, J., Farrow, A., Marín, J. A., Pulido, O. L., Ramírez, A., Gutierrez, T., Sáenz-Cruz, L., Castro, A. and Andersson, M. (2009) *The Andes Basin Focal Project*, CPWF Project Report Series PN63, CGIAR Challenge Program on Water and Food, Colombo, Sri Lanka.

Mulligan, M., Saenz Cruz, L. L. S., Pena-Arancibia, J., Pandey, B., Mahé, G. and Fisher, M. (2012a) 'Water availability and use across the Challenge Program on Water and Food (CPWF) basins', in M. Fisher and S. Cook (eds) *Water, food and poverty in river basins: Defining the limits*, Routledge, London, pp. 255–279.

Mulligan, M., Rubiano, J., Hyman, G., White, D., Garcia, J., Saravia, M., Leon, J. G., Selvaraj, J. J., Guttierez, T. and Saenz-Cruz, L. L. (2012b) 'The Andes basins: Biophysical and developmental diversity in a climate of change', in M. Fisher and S. Cook (eds) *Water, food and poverty in river basins: Defining the limits*, Routledge, London, pp. 9–29.

Mupangwa, W., Walker, S. and Twomlow, S. (2011) 'Start, end and dry spells of the growing season in semi-arid southern Zimbabwe', *Journal of Arid Environments*, vol. 75, no. 11, pp. 1097–1104.

Ogilvie, A., Mahé, G., Ward, J., Serpantié, G., Lemoalle, J., Morand, P., Barbier, B., Diop, A. T., Caron, A., Namarra, R., Kaczan, D., Lukasiewicz, A., Paturel, J.-E., Liénou, G. and Clanet, J.-C. (2012) 'Water, agriculture and poverty in the Niger River basin', in M. Fisher and S. Cook (eds) *Water, food and poverty in river basins: Defining the limits*, Routledge, London, pp. 131–159.

Pearce, F. (2007) *When the rivers run dry: Water—the defining crisis of the twenty-first century*, Beacon Press, Boston MA.

Peden, D., Alemayehu, M., Amede, T., Awulachew, S. B., Faki, H., Haileslassie, A., Herrero, M., Mapezda, E., Mpairwe, D., Musa, M. T., Taddesse, G. and van Breugel, P. (2009) *Nile basin livestock water productivity*, CPWF Project Report Series PN37, CGIAR Challenge Program on Water and Food, Colombo, Sri Lanka.

Pukinskis, I. and Geheb, K. (2012) *The impacts of dams on the fisheries of the Mekong*, State of Knowledge Papers Series SOK 01, CGIAR Challenge Program on Water and Food, Vientiane, Laos.

Quintero, M. (2012) *Project Andes 2: Assessing and anticipating the consequences of introducing benefit sharing mechanisms (BSMs)*, CPWF Six-Monthly Project Reports for April 2012–September 2012, CGIAR Challenge Program on Water and Food, Colombo, Sri Lanka.

Quintero, M., Estrada, R. D., Tapasco, J., Uribe, N., Escobar, G., Mantilla, D., Burbano, J., Beland, E., Gavilanes, C. and Moreno, A. (2012) *Compartiendo los beneficios de los servicios ambientales hidrológicos en la cuenca del Río Quijos (Ecuador)*, Project report, CIAT/CPWF–GIZ–RIMISP, Centro Internacional de Agricultura Tropical, Cali, Colombia and CGIAR Challenge Program on Water and Food, Colombo, Sri Lanka.

Quintero, M., Wunder, S. and Estrada, R. D. (2009) 'For services rendered? Modeling hydrology and livelihoods in Andean payments for environmental services schemes', *Forest Ecology and Management*, vol. 258, no. 9, pp. 1871–1880.

Ratner, B. D., Barman, B., Cohen, P., Mam, K., Nagoli, K. and Allison, E. H. (2012) *Strengthening governance across scales in aquatic agricultural systems*, Working Paper AAS-2012-10, WorldFish Center, Penang, Malaysia.

Rijsberman, F. R. (2004) 'Water scarcity: Fact or fiction?' in T. Fischer, N. Turner, J. Angus, L. McIntyre, M. Robertson, A. Borrel and D. Lloyd (eds) *New Directions for a Diverse Planet*, Proceedings of the 4th International Crop Science Congress, 26 Sep–1 Oct 2004, Brisbane, Australia, cropscience.org.au/icsc2004/plenary/1/1994_rijsbermanf.htm (accessed 10 March 2013).

Rijsberman, F. R. (2006) 'Water scarcity: Fact or fiction?', *Agricultural Water Management*, vol. 80, no. 1, pp. 5–22.

Ringler, C., Cai, X., Wang, J., Ahmed, A., Xue, Y., Xu, Z., Yang, E., Jianshi, Z., Zhu, T. and Cheng, L. (2012) 'Yellow River basin: Living with scarcity', in M. Fisher and S. Cook (eds) *Water, food and poverty in river basins: Defining the limits*, Routledge, London, pp. 218–238.

Rockström, J., Karlberg, L., Wani, S. P., Barron, J., Hatibu, N., Oweis, T., Bruggeman, A., Farahani, J. and Qiang, Z. (2010) 'Managing water in rainfed agriculture—The need for a paradigm shift', *Agricultural Water Management*, vol. 97, no. 4, pp. 543–550.

Rockström, J., Steffen, W., Noone, K., Persson, Å., Chapin, F. S., Lambin, E. F., Lenton, T. M., Scheffer, M., Folke, C. and Schellnhuber, H. J. (2009) 'A safe operating space for humanity', *Nature*, vol. 461, no. 7263, pp. 472–475.

Rubiano J. and Soto, V. (2008) *Extrapolation domains of Project No 16 Aerobic rice system (STAR)*, Internal Report for the IMPACT Module of CPWF, Colombo, Sri Lanka.

Sajor, E. (2012) *Project Mekong 4: Water governance*, CPWF Six-Monthly Project Reports for April 2012–September 2012, CGIAR Challenge Program on Water and Food, Colombo, Sri Lanka.

Scoones, I. (1996) *Hazards and opportunities: Farming livelihoods in dryland Africa. Lessons from Zimbabwe*, Zed Books, London.

Sharifullah, A., Tuong, T. P., Mondal, M. and Franco, D. (2008) 'Increasing crop water productivity through optimizing the use of scarce irrigation water resources', *Proceedings of the CGIAR Challenge Program on Water and Food 2nd International Forum on Water and Food, Addis Ababa, Ethiopia, 10–14 November 2008*, CGIAR Challenge Program on Water and Food, Colombo, Sri Lanka.

Sharma, B. R., Amarasinghe, U. A. and Ambili, G. (eds) (2010*) Tackling the water and food crisis in South Asia: Insights from the Indo-Gangetic Basin*, CPWF Project Report Series PN60, CGIAR Challenge Program on Water and Food, Colombo, Sri Lanka.

Sharma, B., Amarasinghe, U., Xueliang, C., de Condappa, D., Shah, T., Mukherji, A., Bharati, L., Ambili, G., Qureshi, A. and Pant D. (2012) 'The Indus and the Ganges: River basins under extreme pressure', in M. Fisher and S. Cook (eds) *Water, food and poverty in river basins: Defining the limits*, Routledge, London, pp. 30–58.

Sheriff, N., Joffre, O., Hong, M. C., Barman, B., Haque, A. B. M., Rahman, F., Zhu, J., van Nguyen, H., Russell, A., van Brakel, M., Valmonte-Santos, R., Werthmann, C. and Kodio, A. (2010) *Community-based fish culture in seasonal floodplains and irrigation systems*, CPWF Project Report Series PN35, CGIAR Challenge Program on Water and Food, Colombo, Sri Lanka.

Siambi, M. (2011) *Increased food security and income in the Limpopo Basin through integrated crop, water and soil fertility options and public–private partnerships*, CPWF Project Report Series PN01, CGIAR Challenge Program on Water and Food, Colombo, Sri Lanka.

Srivastava, N., Vadez, V., Upadhyaya, H. D. and Saxena, K. B. (2006) 'Screening for intra and inter specific variability for salinity tolerance in Pigeonpea (*Cajanus cajan*) and its related wild species', *Journal of Semi-Arid Tropical Agricultural Research*, vol. 2, no. 1, pp. 1–12.

Sullivan, A. and Sibanda, L. M. (2012) 'Vulnerable populations, unreliable water and low water productivity: A role for institutions in the Limpopo Basin', in M. Fisher and S. Cook (eds) *Water, food and poverty in river basins: Defining the limits*, Routledge, London, pp. 82–109.

Terrasson, I., Fisher, M. J., Andah, W. and Lemoalle, J. (2009) 'Yields and water productivity of rainfed grain crops in the Volta Basin, West Africa', *Water International*, vol. 34, no. 1, pp. 104–118.

Tuong, T. P. and Hoanh, C. T. (2009). *Managing water and land resources for sustainable livelihoods at the interface between fresh and saline water environments in Vietnam and Bangladesh*, CPWF Project Report Series PN10, CGIAR Challenge Program on Water and Food, Colombo, Sri Lanka.

UNDP (2006) *Human development report 2006—Beyond scarcity: Power, poverty and the global water crisis*, Palgrave Macmillan, New York.

Vadez, V., Krishnamurthy, L., Serraj, R., Gaur, P. M., Upadhyaya, H. D., Hoisington, D. A., Varshney, R. K., Turner, N. C. and Siddique, K. H. M. (2007) 'Large variation in salinity tolerance in chickpea is explained by differences in sensitivity at the reproductive stage', *Field Crops Research*, vol. 104, no. 1, pp. 123–129.

van Koppen, B. (2010) *Water rights in informal economies in the Limpopo and Volta Basins*, CPWF Project Report Series PN66, CGIAR Challenge Program on Water and Food, Colombo, Sri Lanka.

Vidal, A. and Perrier, A. (1990) 'Irrigation monitoring by following the water balance from NOAA-AVHRR thermal infrared data', *IEEE Transactions on Geoscience and Remote Sensing*, vol. 28, no. 5, pp. 949–954, doi:10.1109/36.58984.

Ward, P., Rasaenan, T., Meynell, P. J., Ketelsen, T., Sioudom, K. and Carew-Reid, J. (2012) 'Flood control challenges for large hydroelectric reservoirs: Example from Nam Theun-Nam Kading Basin in Lao PDR', in *2nd Mekong Forum on Water, Food and Energy, Hanoi, Vietnam, 13–15 November 2012*. 50.87.54.166/wp-content/uploads/Forum-proceedings.pdf.

World Bank (2007) *World development report 2008: Agriculture for development*. World Bank, Washington DC.

Ziv, G., Baran, E., Nam, S., Rodríguez-Iturbe, I. and Levin, S. A. (2012) 'Trading off fish biodiversity, food security, and hydropower in the Mekong River basin', *Proceedings of the National Academy of Sciences*, vol. 109, no. 15, pp. 5609–5614, pnas.org/cgi/doi/10.1073/pnas.1201423109.

3   Harnessing research for development to tackle wicked problems

Michael Victor,[a][*] *Boru Douthwaite,*[b] *Tonya Schuetz,*[c] *Amanda Harding,*[d] *Larry W. Harrington*[e] and *Olufunke Cofie*[f]

[a]CGIAR Research Program on Water, Land and Ecosystems WLE, Vientiane, Lao PDR; [b]WorldFish Center, Penang, Malaysia; [c]CGIAR Challenge Program on Water and Food CPWF, Munich, Germany; [d]CGIAR Challenge Program on Water and Food CPWF, Paris, France; [e]CGIAR Challenge Program on Water and Food CPWF, Ithaca, NY, USA; [f]International Water Management Institute IWMI, Accra, Ghana; [*]Corresponding author, M.Victor@cgiar.org.

Introduction: From a linear model to an iterative approach

The CGIAR Challenge Program on Water and Food (CPWF) saw water management as an entry point for addressing broader development objectives, which, although complex, are usually closely linked. The primary objective was to produce more food while maintaining the sustainability and resilience of the agroecosystems. This meant using technical, institutional and policy change to improve water management to reduce poverty, ensure equitable resource use, preserve ecosystems services and adapt to climate change.

These are wicked problems (Rittel and Webber, 1973), which are difficult to solve because their requirements are contradictory, changing, hard to reconcile and often not well understood. Solutions typically require many people to change mindsets and behavior. Wicked problems require an integrated social approach to consult and negotiate common understanding as a problem unfolds so that responses to it are acceptable to different social groups whose interests may conflict (Carlile et al., 2013).

Wicked problems are not amenable to traditional linear model approaches where innovations are primarily technical, such as new technologies to improve productivity. In the linear model research results and technologies flow from research stations into farmers' fields.

The CPWF underwent a process of learning how to carry out research on wicked problems that is relevant and related to users' needs, including farmers, development agencies and policy-makers. In its Phase 2 (2009–2013) the CPWF adopted a research-for-development (R4D) approach, which evolved with experience (Hall, 2013).

In this chapter we describe what the CPWF learned about R4D and addressing wicked problems, which we will refer to as development challenges. We will introduce basic concepts, principles and definitions of R4D as the CPWF came to understand them, and describe how it applied them using examples from CPWF projects in basins. In the process, we will include an analysis of the CPWF experience of putting theory into practice and draw out some guiding principles.

Concepts and principles of research for development

There are a number of terms related to what CPWF came to think of as R4D. These include, amongst others, research for development itself, participatory technology development, research in development, integrated agricultural research for development, and social learning. R4D is a set of approaches that may lead to innovation under differing circumstances. Some critics argue, however, that, "what . . . [R4D] needs is . . . some practical guidance on which ways of using research for innovation should be supported and under which circumstances" (Hall et al., 2010).

One way R4D differs from the linear model of research and extension is by encouraging many people to take part in the research process. R4D uses a systems perspective to bring people together and to illustrate how many development challenges are interlinked across multiple scales (Hawkins et al., 2009).

R4D has many roots. A partial list includes applied and action research from social anthropology 1930s–1950s; farming systems research 1970s–1990s (Byerlee et al., 1982); agroecology (Conway, 1986); on-farm client-oriented research (Tripp et al., 1990); participatory research and participatory action research (Whyte et al., 1989); integrated natural resources management research (Campbell and Sayer, 2003); and ecoregional research (Harrington and Hobbs, 2009). By the early 2000s, there was growing criticism of traditional linear research models and calls for more client-driven and inclusive research approaches. The Forum for Agricultural Research in Africa commissioned a benchmark study on Integrated Agriculture Research for Development which set out basic principles for R4D (Hawkins et al., 2009; Hall et al, 2010; Ugbe, 2010):

- Generating innovation amongst stakeholders rather than focusing solely on research products and technologies.
- Embedding and interdependence of innovation processes in local institutional, policy and political contexts.
- Defining development objectives in a given area and using a flexible approach.
- User involvement and continuous interaction with stakeholders so that their knowledge, perspectives and needs are integrated into the research process.

- Analysis and incorporation of lessons learned.
- Research is not the center of the work itself but is a distributed activity involving a dense network of people.

Innovation is a key component of R4D, which is defined as "novel ways to do things better." It is about the "how" of doing things differently that triggers change, not just the "what" of new outputs, although they also are important (Perrin, 2002). An innovation system is a network of organizations and people, together with the policies that affect their innovative behavior. The system brings new processes, new forms of organization and new products into economic use. Innovation is often driven by entrepreneurs pursuing market opportunities (Hall et al., 2010).

The most important tool in R4D is the "learning system." Learning systems include approaches such as learning by participatory action or social learning. Both focus on a series of steps that help groups of people learn while they are doing (Mbabu and Hall, 2012). Learning systems feature:

- an effective performance management or monitoring and evaluation system;
- facilitated strategic and organizational planning;
- a research culture that supports an institutional learning and change agenda;
- research operations that explore processes and pathways to impact; and
- links to communities of practice sharing experiences and lessons.

Harnessing complexity: Theoretical basis of the CPWF approach

In its evolving R4D program, the CPWF built on the networks of water and agricultural scientists of Phase 1. It developed solutions to local problems by involving the people who would adopt those solutions. In 2007, the CPWF applied one variation of the theory of change (ToC) (Weiss, 1995) to its new R4D approach. It asked, "How can the small investments that the CPWF can make have an impact on the lives of the millions of people living in the basins in which it works?" This led the CPWF to the concept of "emergence," wherein complex systems and patterns arise out of a multiplicity of relatively simple interactions.[1] Hence, the CPWF's ToC is to carry out research that fosters emergence (Box 3.1).

The ToC the CPWF adopted is based on: (1) the theory of complex adaptive systems, coupled with learning selection; (2) social network theory; and (3) program theory (Rogers et al., 2000). We describe how each of these bases contributes to our ToC in terms of "cornerstones."

Box 3.1 Components of CPWF's ToC

1   Improvements in water management and water productivity come from both social and technical innovation.
2   Innovation systems in basins and sub-basins are complex adaptive systems.
3   Grassroots innovation processes in complex adaptive systems are driven by "learning selection."
4   Components 2 and 3 provide a framework for problem analysis to devise research strategies.
5   The CPWF carries out its research to develop strategies and interventions, through strengthening networks.

Source: Douthwaite (2011)

Cornerstone 1: Understanding systemic change

In its second phase, the CPWF conducted R4D on particular Basin Development Challenges (BDCs) in each of six river basins. The use of BDCs allowed the CPWF to focus on specific problem sets in the context of broader complex adaptive systems found in basins. (See Chapter 4 for discussion of BDCs and how they were prioritized and selected.) In each basin, the CPWF conducted research on BDCs through sets of interlinked projects with the aim of developing systemic understanding and triggering "tipping points" that would lead to innovation and change.

Emergence is an inherent property of complex adaptive systems (Axelrod and Cohen, 2000). Emergence results from the interactions between agents (people in our case), strategies (what to do in which circumstances), and artefacts (material resources that respond to the action of agents). Emergence is not a property of any single agent, nor can it be predicted easily or deduced from the behavior of individual agents. There are many examples of emergence from many different types of complex adaptive systems. For example, the constantly changing shape of a flock of birds in flight or the construction and maintenance of a termite mound occurs because individual birds or termites have a few simple genetically coded rules that they follow.

In complex adaptive systems in human communities, emergence is driven by learning selection (Douthwaite, 2002). People experiment by trying novel ways to do things. If they succeed, they may decide to continue with the novelty, adapt it, or abandon it. While they experiment, they interact with others, who may influence what they decide to do with the novelty. This is the process of learning selection.

Learning selection applies to institutional as well as technical innovations. In responding to problems of collective water and food management, institutions "evolve through complex creative processes that adopt and adapt diverse ingredients" (Merrey and Cook, 2012). Merrey and Cook's concept of *"bricolage"*[2] is useful to promote and facilitate a creative approach to building and strengthening water management institutions.

Learning selection by large numbers of people linked together often produces innovation. The process is spontaneous, although it can be nurtured by facilitators and guided by product champions who play distinct but important roles. Champions of learning selection are more effective where knowledge and experience are freely shared. The outcome of learning selection depends on people's motivations and their ability to participate, which depends in turn on a community's power relations and cultural norms.

Based on this analysis, research can foster large-scale change in the CPWF basins by understanding and stimulating:

• varying types of agents, strategies and artefacts, for example by introducing novelty into the system, such as a new crop variety or other technology, and fostering its local adaptation;
• changes to patterns of interaction between agents, strategies and artefacts, for example changing social norms; and
• changes to selection processes by which the fitness of an agent, strategy or artefact is assessed, and the subsequent processes that allow those judged to be fitter to survive and spread.

Learning selection can foster large-scale change, but not necessarily change that is equitable or sustainable. There is a role for research to anticipate and monitor the consequences of change in production systems. See Chapter 5 for further discussion on the consequences of change.

Cornerstone 2: Network weaving to foster emergence

Improved connectivity between actors relies on improved understanding and mapping of networks. The CPWF used the concepts of social network analysis (Cross and Parker, 2004), small world networks (Watts and Strogatz, 1998) and network weaving (Krebs and Holley, 2002) to analyze and strengthen its networks. For example, a map of a research network could show the organizations involved and how they are connected. Changing the ways in which organizations and individuals are connected changes who interacts with whom, how ideas spread and how decisions are made.

Figure 3.1 illustrates the concept of network weaving as applied to a hypothetical river basin. Initially research takes place among scattered clusters of individuals and organizations who do not know what the others are doing (Figure 3.1a). This network does not have critical mass, learning selection is slow, and the chances of innovation are low. We now add a network weaver, who can be one or more individuals or organizations, to link and foster communication between the separate groups and create a hub-and-spoke

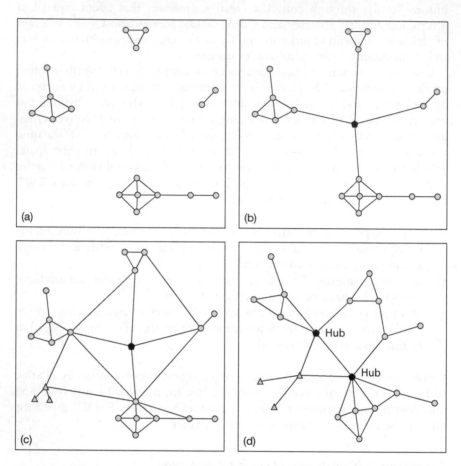

Figure 3.1 The network weaving model the CPWF used for building innovative small
world networks of food and water research and development: a) scattered
clusters; b) hub-and-spoke network; c) multi-hub small world network; and
d) same multi-hub network with the network weaver removed.

Source: Authors' original data.

network (Figure 3.1b). This is better, but the structure is unstable because if the
network weaver (pentagon) leaves, the network reverts to scattered clusters.

Now we add new links to people and projects in other basins (triangles,
Figure 3.1c). The structure is much more robust and can withstand losing the
weaver (Figure 3.1d). The effect of the weaver changed interaction patterns to
a network from which innovations are much more likely to emerge. Such
resilience was demonstrated in the closing of CPWF. As discussed in Chapter
4, the CPWF went through a transformation due to CGIAR reform. The
CGIAR Research Program on Water, Land and Ecosystems (WLE) into which

the CPWF is being integrated is developing its own focal region programs and in doing so is drawing on the CPWF experience.

The Andes provides another example of robust networks. One project had good relations with the Minister for Environment and Natural Resources in Peru, but realized that ministers have uncertain tenure. The team built relationships with others in the Ministry to ensure that the project could continue to influence policy (Quintero, 2012).

In principle, network structures of clusters linked by a few hubs are very effective, resembling small world networks, in which most pairs of nodes are connected by at least one short path. They feature: (1) clusters that form around common attributes and goals, which are needed for trust, and many connections and nodes, which maximize innovation; and (2) short average path lengths without requiring that each node is directly connected to all others. The CPWF created network links by funding and facilitating multi-stakeholder projects through regional and global fora, and by training. The CPWF did not create new network structures, but created links among existing networks. The CPWF learned to "build on what is already there rather than set up new platforms and systems" (see Chapter 7).

Cornerstone 3: Program evaluation

The CPWF required each project to show how its research would improve the livelihoods of its target group and how that would happen. This was called the project's outcome logic model (OLM). Regular monitoring and evaluation (M&E) determined how well each project met its OLM. In practice it was difficult to evaluate OLMs within M&E. Both were de-emphasized as a result of the 2012 funding cuts (see Chapter 4).

Cornerstone 4: Building the capacity to work differently

R4D was new for the CPWF and its partners so that training and mentoring evolved together with good knowledge management. A key component was product championing with researchers taking responsibility for the impact of their research. Training and knowledge management were powerful ways to build connections between people and organizations.

Lessons from CPWF's approach to R4D

We divide the CPWF's research into three categories. Phase 1 projects explored a wide range of concepts, procedures, innovations and partnerships. Basin Focal Projects (BFPs) revised CPWF assumptions on water availability, scarcity, water productivity, water and poverty, and water-related institutions (see Chapter 2). Subsequently, Phase 2 projects were designed around R4D with projects grouped into basin-level programs. Each project focused on a BDC.[3] Each basin program consisted of three or four technical projects with a coordination project led by a Basin Leader.[4] The Mekong Basin was an exception with funding from AusAID allowing 19 projects in five countries.

R4D in Phase 2 was based on the assumption that innovation is essential for sustainable change to occur. Social and economic change as described in OLMs arose from technological and institutional innovations fostered by interactions among stakeholders and partners. The CPWF contributed to learning selection in the innovation process by training people to work differently and linking behavior to outcomes specified in the OLMs.

The CPWF R4D framework and processes allowed researchers in basin programs to understand the complexity of agricultural water management, and how scale affects the impacts of research. Basin teams discussed with stakeholders about how change takes place and how best to support it (Hall, 2013). The form of the discussions differed between basins, but they were guided by a CPWF-wide emphasis on adaptive management, reflection and learning. They took the form of stakeholder meetings at the research design stage, annual reflection meetings in the basins, and engagement platforms at local, national and regional level.

In 2013, CPWF Basin Leaders and Project Leaders began a process to learn and reflect on the diversity of approaches applied under R4D (CPWF, 2013). Thus the definition and framework described below is inductive and ex-post. We use examples from Phase 2 BDCs to explain the framework and how it evolved.

R4D is defined by the CPWF as: "An engagement process for understanding and addressing development challenges defined with stakeholders. Stakeholders are champions and partners in the research process as well as the change it aims to bring about." CPWF's practical experience of R4D in BDCs is distilled into eight principles (Box 3.2).

We now discuss sub-sets and combinations of these principles.

Box 3.2 The eight principles of the CPWF's R4D

1 Theory of change;
2 knowledge management;
3 partnerships and networking;
4 research on R4D;
5 policy and engagement;
6 adaptive management;
7 gender and diversity; and
8 capacity building.

Source: CPWF (2013)

ToCs and impact pathways

> To define a [ToC] and to follow the evolution of potential innovations through outcome pathways—as requested by the CPWF—appears to be more than a conceptual and analytical tool: this is now an order imposed by our end users.
>
> (Chechi, 2012)

All CPWF BDCs used the ToC. It was applied through M&E and learning tools. Basins varied in the way they used the ToC, which has evolved over time. The CPWF describes its approach to ToC in its monitoring and evaluation guidelines (Alvarez et al., 2010) as:

> A [ToC] is the causal (or cause–effect) logic that links research activities to the desired changes in the [people] that a project or program is targeting to change. It describes the tactics and strategies, including working through partnerships and networks thought necessary to achieve the desired changes in the target [people]. A theory of change provides a model of how a project or a program is supposed to work . . . The value of testing and refining the model is that it challenges preconceptions, aids reflection and helps staff ask themselves, "Are we doing the right thing to achieve the changes we want to see?" Regularly asking this question, and responding to the answer is essential good practice for any [R4D] project or program.

The CPWF made ToC operational using a number of tools, including participatory impact pathways analysis (PIPA), which is a method developed for planning complex projects to achieve outcomes. It uses a participatory process to promote learning and provide a framework for research on change (Douthwaite et al., 2007). PIPA was applied to all projects at the initial stages in Phase 2.

In turn, CPWF then applied OLMs to describe projects' medium-term objectives for translating research outputs to behavioral outcomes (Box 3.3).

Box 3.3 **Components of an OLM**

- What changes in behavior (policy or practice) are needed to achieve outcomes?
- What changes in knowledge, attitudes or skills are needed to change behavior?
- Which people need to change policy or practice to get to outcomes?
- What strategies are needed to influence knowledge, attitudes and skills of these people?
- How can research outputs be used and leveraged within these strategies?

As a model of possibilities, PIPA makes explicit: (i) the causal links between outputs, outcomes and impacts; and (ii) the relationships between partner organizations that are necessary for these to happen. Impact narratives, as articulated in the OLMs, improve stakeholders' understanding and communication. In principle, impact pathways can address incentives, power differentials and cultural values.

The CPWF required all Phase 2 projects to use a ToC, mostly by discussing the desired outcomes, developing OLMs and linking them with project activities. The CPWF used M&E, linked to ToC, on projects' impact pathways. Methods included most significant change stories (Leon et al., 2009; CPWF, 2012) and reflection workshops, and progress and annual reports.

Basin teams recognized that using ToC, including PIPA, was useful to share a project's vision, and to analyze entry points and progress toward outcomes. This is a distinguishing feature of the CPWF. That said, it was also reported that applying many of the tools developed for project monitoring and compliance were laborious and time-consuming. Many also had difficulty understanding how individual projects fit within basin and global programs.

Some now think that the CPWF should have put more emphasis at the outset on the theory and practice of R4D (Cofie, 2012). The most obvious gaps were researchers' reluctance to engage stakeholders in agenda setting, research design and monitoring (Sullivan, 2012). Basin project team leaders provided a number of critical thoughts (Box 3.4).

When Phase 2 started, the CPWF proposition was that the R4D approach would trigger tipping points for large-scale change over the ten years of Phases

---

Box 3.4 Thoughts from basin and project team leaders

- "Outputs to outcomes to impact" is NOT a common language among researchers, many of whom are reluctant to consider targeting research outputs to outcomes.
- Cross-project, cross-discipline, and cross-scale linking of milestones to communications plans is a more robust approach than using individual projects, disciplines and scales.
- OLMs should be conceived as iterative and not a snapshot. Outputs evolve over time and opportunities to link them to outcomes may bring in new people.
- Sharing expected outputs with a wide range of trusted audiences throughout the process increases the likelihood of getting the right information into the right hands at the right time.
- Some outputs intended for internal use only (a data layer, for example) can be useful to other people, if they are made aware of them. Conversely, some outputs may be considered sufficient if they contribute to the global body of knowledge.

2 and 3. When Phase 3 was canceled, projects were reduced to three or four years (excluding set up and launch). Some projects (see Chapter 8) are nonetheless close to tipping points. "Change tends to be gradual and [imperceptible], obtained through close and constant attention to relations between the [BDC] and its target groups" (Geheb, 2012).

Partnerships, networking and engagement

The CPWF network is probably too expensive a model if your goals are peer reviewed journal articles yet if the goal is broader, say R4D or development of a specialized team, then the calculations would be different. Partnerships get less expensive the longer they last and the more reliable they become.

(Sullivan, 2012)

Partnerships and networks were central to the CPWF model of R4D. Basin Coordinators and Project Leaders focused on developing partnerships and collaboration. They found that personal contacts, social capital and engaging in networks were key ways to advance R4D. The CPWF identified several categories of partners and partnerships (Box 3.5).

A unique characteristic of the BDCs was that partners other than from CGIAR Centers led 14 of the 29 commissioned projects, with more regional and national partners. The large Mekong Program, with 19 projects, had 76 contracted partners and memoranda of understanding with seven government agencies in the region (Clayton, 2013). The memoranda provided access for CPWF field teams, formalized relationships between government agencies and the program, and spun off new relationships and new initiatives (Geheb, 2011).

Key features of the CPWF R4D model were its roles as convener, engager, negotiator, enabler, space provider and trusted broker. With CPWF research outputs intended to contribute to development outcomes, partners brought quality research to policymakers' attention. They also promoted conversations between key people in research, policy and development. R4D required partners with clear links, roles, mandates and perspectives, which provided research-based, informed decision-making. The BDCs of Phase 2 focused on

---

Box 3.5 Categories of CPWF partners and partnerships

- Hosting and convening arrangements.
- Implementation partners, contracted partners who actually implemented the agreed-upon program of work.
- Next users/end users.
- End users who we were trying to change. CPWF engaged directly with its end users to make sure they were part of the research.

local realities and people, national policy frameworks and included relevant, knowledgeable partners. In the following sections we reflect upon partnerships, networks and engagement.

Across the basins, the project teams learned the importance of engaging the intended "targets" of the research from the outset, which blurred the boundaries between researchers and users. This ensured that the research was owned and used by different users.

Quintero (2012) identified a number of principles for engaging different types of partners:

- Share resources and capacities: The project achieved agreements with users who provided human and sometimes financial resources to implement agreed activities.
- Define the research agenda jointly, if necessary explaining to partners simply and understandably the required methods and approaches.
- Ensure that partners understand the research results.
- Hold all public meetings jointly with partners, emphasizing the inter-institutional collaboration toward common objectives.
- Let others lead the process of creating a benefit-sharing mechanism (BSM), while providing support to their arguments and BSM designs with research results.
- Be flexible toward changing original plans; in some cases the orientation of analysis has to change. For example, in some cases it was more important to value the benefits already produced by ecosystem services than assume the opportunity cost of changing the supply level of the services.

Trust

Fruitful partnerships are built on trust, but developing social capital and gaining partners' trust takes time. For example, the Mekong BDC focused on sustainable hydropower. This is contentious and highly political with many conflicting interests (Lebel et al., 2010). The Mekong BDC achieved the status of trusted broker by providing neutral ground where dialogue could take place. It manifested trust, exemplified by the annual high-level Mekong Forum convened by the BDC (Geheb, 2012).

Common vision

R4D progresses when partners share a common vision of a development challenge and the ways to address it. A common vision means a common vocabulary, which helps each partner see their role and how they might contribute. The vision may evolve through learning selection. The program vision should be part of the ToC.

The Limpopo BDC showed that a common vision that all partners can accept and follow could be flexible by allowing for change as the program developed. It also allowed strong leaders and team members to be used to their fullest. Nevertheless, individual project team accomplishments within a complex partnership must be recognized, not hidden (Sullivan, 2012).

In the Andes BDC, a strong common vision allowed other partners, including national institutions, to lead innovation and change, with the CPWF offering support through relevant research results (Quintero, 2012; Saravia, 2012).

Understanding mandates and making strategic alliances

It is important that partnerships include institutions or organizations with authority, decision-making power and credibility. Not including them jeopardizes the R4D program.

The Mekong BDC found that formal memoranda of understanding with the appropriate national authorities enabled progress.

> The memoranda of understanding have served to gain access to governments; they have worked to reduce ambiguities and mistrust. They have provided access for our field teams. They have formalized relationships between the state and the program. And they have served to spin off new relationships and new initiatives.
>
> (Geheb, 2012)

Further to making strategic alliances, we again emphasize the need to be flexible, even allowing the direction of the project to change. As we stated above, it may be better to take account of the value obtained by using one or more ecosystem services than to pay the cost of conserving them (Quintero, 2012).

The R4D design process

Projects will have a greater likelihood of influencing policy if they engage national and regional level decision-makers at the beginning of the R4D process. Keeping them engaged throughout the project ensures relevance and shares ownership.

In the Ganges BDC, early engagement with decision-makers helped to align projects with national priorities and policy—for example, government policies on maintenance of rural infrastructure, which was a national priority (George and Meisner, 2013), were the key to control water for farm intensification and diversification proposed by the BDC.

In the Nile, the Ethiopian national sustainable land management (SLM) program was involved in the BDC's innovation platform and steering committee and therefore in the stakeholder consultations. The Nile BDC was not engaged with the implementation of the SLM program, although it helped to reveal its top-down nature. The Nile BDC's key messages were strongly endorsed by the Ethiopian stakeholders, including policy-makers (Merrey et al., 2013).

Convening power

A basin R4D program is best managed by an organization with convening power. CPWF teams contracted regional institutions that had convening power to lead the engagement effort: the Food, Agriculture and Natural Resources Policy Analysis Network (FANRPAN, Limpopo), Consortio para del Desarrollo Sonstenible de la Ecoregión Andina (CONDESAN, Andes) and Volta Basin Authority (Volta). In the Ganges, leaders were local partners (Bangladesh Rural Advancement Committee).

The organization itself may have convening power (FANRPAN and CONDESAN) or a new platform might be necessary. In the Mekong, the CPWF Mekong coordination and change project became a trusted convener of hydropower stakeholders (investors, governments, NGOs and development agencies) in a contentious field of work (WLE, 2013).

Convening power is based on social capital including trust, credibility, relevance and mandate. It is most effective when combined with easy access to high-level decision-makers. But developing social capital takes time and is vulnerable to external shocks, for example, unanticipated budget cuts or national or regional policy adjustments. Convening power has been consistent across all six CPWF basins, increasing over time. As the CPWF shared its research results with partners, it established and defined common interests better. It was then able to engage with complex policy-level issues.

Complexity and adaptive management

Social development and policy change are long-term, non-linear processes, full of short-term decisions that directly influence progress. It is therefore important to be flexible and to change plans if the circumstances require it.

All BDCs have examples of complexity and adaptive management. The Nile BDC learned that successful landscape management required reconciling top-down national priorities for soil and water conservation with community needs (Merrey and Gebreselassie, 2011; Ludi et al., 2013). The Ganges BDC adjusted its research questions several times as they better understood the complex interrelationships between system intensification, the component technologies of the farming systems, the coordination and timing of water control, the design, repair and management of rural infrastructure, and the overlap between national and local government policies and priorities (George and Meisner, 2013). The Limpopo BDC built on past achievements but also found new ways forward for the design of water infrastructure for multiple uses, and market development for small livestock (van Koppen et al., 2009; van Rooyen and Homann-Kee Tui, 2009; Sullivan, 2012). In the process, partners adjusted to how they perceived and addressed opportunities.

Timelines and dilemmas

The time-bound nature of the CPWF complicated its engagement in policy processes and its legitimacy to do research and enable change. Decision-making is a long process, vulnerable to multiple influences and driven by personalities. Who can take credit for which outcomes is therefore problematic and not always relevant.

Local partners will continue to engage in the change processes as long as there are sufficiently high levels of trust, credibility and legitimacy. Credible and relevant research used by stakeholders in effective R4D increases the likelihood—but does not ensure—that outcomes will be achieved.

CPWF anticipated these vulnerabilities by ensuring that the BDC's coordination and change programs were either led by local institutions or had strong partnerships. In some cases, outputs will likely be translated to outcomes after the CPWF ends (see Chapters 5 and 6 for more on engagement in policy processes).

Inclusive and participatory communication

> [C]ommunications is not just one element in the struggle to make science relevant. It is the central element. Because if you gather scientific knowledge but are unable to convey it to others in a correct and compelling form, you might as well not even have bothered to gather the information.
> (Olsen, 2009)

Conventionally, communication has been perceived as something to do at the end of the research cycle. Research produces results, which are disseminated as a poster, a manual, a policy brief, or a glossy brochure for donors. But this approach has not proven to be up to the task (PANOS, 2007) in ensuring that communication processes support moving research results to outcomes. Each basin took a strategic approach to communications, which focused not just on the products but on processes for engagement.

In the Volta one project implemented an explicit communications strategy that combined:

- One-on-one interviews with key stakeholders, such as the local administration, to engage them in the full R4D process.
- A consultation process with groups including presentations of the project at the District Assembly in Ghana and at the Plan d'Action pour la Gestion Intégrée des Ressources en Eau (PAGIRE, the official program to support integrated water resources management (IWRM) in Burkina Faso), workshops with stakeholders at various levels at the community, district and regional levels in Ghana (Kizito, 2012).
- Establishment of a group of experts from administration, researchers, NGOs and universities, all experienced in IWRM. The aim was to help build the strategy for the participatory process.

In the Mekong, the BDC communications strategy (Geheb, 2010) described the R4D process and its results and was elaborated as the program progressed and grew. The communications strategy was a key element given the Mekong's complex power relations, the research focus and its contribution to the Basin's development. The strategy was based on an analysis of the policy and statutes, stakeholders, power relations and establishing partnerships at all levels. It used innovative products and processes such as a series of documentary films, which successfully stretched the limits of public dialogue (Clayton, 2013). It also used protocols such as the Hydropower Sustainability Assessment Protocol to bring actors together for dialogue.

Research on hydropower plants in the Colombian Andes showed that nearby communities were disempowered with no share in the benefits that the plants generated. There was little communication with the local population and policy-makers were inadequately informed. The research reported the communication breakdown as a challenge to link the up-stream and down-stream components of decision-making. It created contacts between decision-makers and local stakeholders, to ensure that local concerns were considered (Mulligan, 2013).

Gender, diversity and power

Power relations influenced the research process in all basins (see Chapter 7). In the Andes, CPWF's partnership with the World Wildlife Fund (WWF) enabled a citizen participatory call to action (*conversatorio*) for accountability and improved management of natural resources. WWF facilitated the *conversatorios*, which improved local knowledge and capacity to negotiate conflicts over access to water and equitable distribution of common goods. Women played an important role, using existing legislation on participatory decision-making mechanisms to ensure their voices were heard (Candelo et al., 2008; Córdoba and White, 2011). The CPWF adapted information from different modeling scenarios to level the playing field between investors, government agencies and local communities (Mulligan, 2013).

Integrating gender in R4D

Women play a critical role in agriculture in developing countries. In sub-Saharan Africa almost 50 percent of rainfed smallholders are women, producing 70–80 percent of domestic food in most societies (Gladwin, 2002; FAO, 2013). Despite this, in many instances, women cannot own land or control social and economic resources. Women often cannot be members of rural organizations and cooperatives, and often do not have access to agricultural inputs and technology such as improved seeds, training, extension and marketing services. It is harder for rural women than it is for men to secure their livelihoods, which depend on agriculture. Inequity in gender limits a people's economic and social development. It is therefore important that the CPWF's R4D took account of gender.

Gender roles shift with social, economic and technological change. For example, the introduction of new crops and technologies, mounting pressure on land, or increasing poverty or migration can change the roles of men and women in agriculture. Migration can empower the women who stay in rural areas when men leave to work in towns. When technical approaches incorporate gender concerns, they can empower women, but when they do not the consequences are likely to be negative for women and girls.

Gender was included in Phase 2 research design and capacity and the learning mechanisms. Initially research proposals included gender but only in a marginal way. Then the CPWF adopted an explicit program-wide gender initiative (2011–13) (Box 3.6). This was based on a multi-pronged approach to transform researchers and R4D by asking the question, "Are there barriers to women's full participation in any activity—workshop, adoption, new technology?"

The focus on gender led to data disaggregation, adaptation in action research and more targeted use of participatory approaches. For example, this included specific consideration of men's and women's differing roles in crop and livestock management from the Nile BDC. It found that women farmers' incomes increased and that women had an increased role in decision-making when involved in differentiated ways in local innovation platforms. In Zimbabwe, where women represent nearly half of all goat owners, an innovation platform changed the goat market from on-farm purchase by intermediaries to formal public auctions. This changed the power dynamics in favor of women through stabilizing prices, rewarding quality, and hence promoting farmer investment and innovation (ICRISAT, 2011).

In the Volta BDC, CPWF partners researched gender dynamics in small reservoir management. They found that women farmers grouped together to overcome local and national barriers to access financing, technology and agriculture-based decision-making (Lasiter and Stawicki, 2013).

Lessons learned include:

1   Integrate rather than isolate by ensuring gender is taken account of throughout projects: in survey design, innovation platforms, M&E systems

---

Box 3.6 Components of the CPWF new gender initiative

- Basin gender audits;
- gender checklist;
- basin gender awareness training;
- revision of monitoring and evaluation systems;
- commissioning gender-specific research;
- special sessions in IFWF3;
- gender stories; and
- a sustained gender conversation across CPWF project teams.

Source: CPWF Website (2013)

rather than as isolated gender projects. Ideally gender should be integrated at the design stage.

2 Move from discourse to practical action with demonstrable gender results. This requires leadership, resources, consistency, incentives and vision balanced with realistic ambition.

3 Value multiple strategies that recognize and demonstrate the centrality of gender in the research and the necessity of individual and organizational cultural change.

4 Share the expertise through companion science where social scientists/gender experts within teams spend time with and accompany engineers and modelers.

5 Gender is increasingly seen within wider considerations of power and alongside dynamics of youth, indigenous people and religious differences.

6 Concrete documentation emerging from research should cover both research and the process of gender integration.

7 Addressing gender requires multiple approaches; there is no magic bullet.

8 By understanding the many forms of gender attitudes, the CPWF community can address the issues of power and voice, which the poorest and most vulnerable lack.

9 R4D that is relevant and credible must include gender and equity so that women can access resources and engage in the development process.

10 Examine attitudes to gender in implementing institutions so that researchers and implementers understand their role in addressing gender inequalities and are willing to change themselves.

11 Planning and gender responsive learning (monitoring, review and evaluation) requires finance and skilled technical human resources.

Knowledge management and communication

Knowledge management

The CPWF defined knowledge management (KM) as the process of capturing, developing, sharing and effectively using knowledge. KM and communications are integral parts of the research process and a field of research in its own right (Harvey et al., 2012).

The CPWF saw KM as central to learning and innovation (Box 3.7), concerned with managing knowledge produced by research to influence decision-makers. It was central to stakeholder engagement, networking and partnerships and was therefore important in achieving CPWF research and development goals (CPWF, 2010). This required three things:

- Having the right information available to support decision-making;
- Using M&E to obtain the right information; and
- Using communications tools to influence knowledge, attitudes and skills in support of behavioral change.

Box 3.7 Factors that drive KM as part of research

- Boundaries between KM and communication, M&E, information management and information communication technology are becoming blurred;
- increasing need to consider the processes by which research products are communicated;
- views differ as to what constitutes "research communication" and how it is done;
- communication is a social and dynamic process rather than a linear one; and
- KM and communication are democratic so that everyone can be involved, which is both an opportunity and a challenge.

Source: CPWF (2011)

Communications and information

In Phase 2 relative to Phase 1, CPWF invested more resources, time and effort into KM and communication, testing how to communicate in ways that supported R4D. In the conventional research process, communication comes at the end of the research process. In R4D, communication is a continuous process that aims to create shared meaning and is a strategic function to achieve outcomes rather than a corporate support function. For the CPWF, communication was an integral part of the R4D process, not just a support function. The key aspects of the CPWF's communication strategy were:

- Communication objectives change with time (Figure 3.2);
- communication is part of the research process and therefore requires new relationships; and
- communication is not linear but a multi-directional, iterative and two-way process.

Internal communications among partners

At the start of Phase 2, program partners needed to share knowledge, gain trust and understand better how different CPWF projects and activities could interact. Each basin team established different mechanisms for internal communication with varying usage and success. Improving internal communication was often a struggle requiring researchers and managers to change their behavior (Merrey et al., 2013).

Table 3.1 lists the communication tools used in the different basins. The tools had mixed reviews. Some senior researchers participated fully, while some

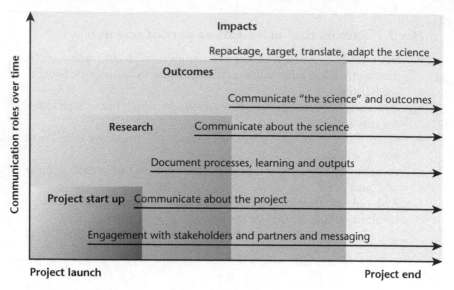

Figure 3.2 Changes in communication roles over time.

Source: Ballantyne, 2012.

national researchers felt marginalized because of poor internet access and unclear user protocols (Merrey et al., 2013). In the Volta, online systems did not work well because of language differences and poor internet coverage.

All basins emphasized regular in-person meetings, focused science meetings and annual or semi-annual reflection workshops to refine working approaches. The Mekong established an annual forum, where researchers presented their results to investors, government officials, development professionals and other researchers. In the Andes and the Ganges, partners used field tours followed by cross-project technical meetings to develop a common vision of the issues and build synergies.

The CPWF used dialogue spaces, online discussion groups, repositories, blogs and a website to improve communication between and amongst basin teams. Nevertheless, a survey in 2013 found that internal communication could have been better. Most researchers believed that the tools were more useful for the coordination and change project team members and Project Leaders. Program information sent through these channels often did not reach project researchers or partners. In contrast, the CPWF website and e-letter were perceived to be very useful for sharing information across basins (Schuetz, 2013).

A challenge throughout the Program was the proliferation of systems that replicated functions. Many basins developed their own systems for sharing, document repositories and even social media (Table 3.1). Some projects developed their own websites and sharing mechanisms.

Table 3.1 List of internal communication tools used within the CPWF Program.

Tool	Andes	Ganges	Limpopo	Mekong	Nile	Volta	Program level
Internal discussions	Email	Email	Email	Mekong list	Yammer	Google Group	Yammer
Blog	Website Project blog	CPWF website	None	Basin website	Basin blog	Basin website	Program website
Internal sharing space	NA	Wikispace	Wikispace	Wikispace	Wikispace	Wikispace	Google sites
Document repository	Own system	CGSpace	Own system	Own system	CGSpace	CGSpace	CGSpace
Website	Own	CPWF website	FANRPAN	Own	Basin blog	Own	Website
Social media	Own	CPWF	None	Own	Own	Own	Own
Workshops and meetings	Reflection workshop	Reflection workshop Study tours	Reflection workshop	Annual forum	Reflection workshop	Reflection workshop	Peer assists IFWF[a]

Source: Authors.

Note: [a]IFWF = International Fora on Water and Food. There were three: in Vientianne, Laos PDR in 2006; in Addis Ababa, Ethiopia in 2008; and in Tshwane, South Africa in 2011.

Research and communications in engagement strategies

The CPWF Communications group worked with researchers to foster engagement with decision-makers to help translate research outputs (information) to development outcomes (changes in decision-maker policy or practice). R4D requires researchers to learn new skills to communicate, including brokering, knowledge sharing and the ability to tailor their presentations to different audiences and different contexts (Table 3.2). The emphasis was to ensure that communication processes were linked to the change identified in impact pathways. R4D communication includes the needs of conventional research, but in R4D, materials are focused on supporting changes identified along the impact pathways. The divisions between "users of research" and participants in the research often become blurred. This is different from carrying out isolated studies and then packaging messages to end users in the expectation they will use and adopt them.

In the examples below we show how the CPWF used communication strategically to enhance research outcomes.

Table 3.2 Different needs for communication in conventional research and in R4D research.

Area	Conventional research communication	Communication in R4D (includes column 2 as well)
Objectives	To inform and provide information	Change perceptions, discourse, policies and behaviors, contribute to development processes
Targets	Researchers, scientists, academics	Multiple actors (farmers, planners, policymakers, private sector, NGOs)
Methodology	One-way Passive	Two-way; multiple actors involved Participatory Engaged and active
Strategies	Publish in journals Attend scientific meetings Message-focused Hand over information to media, public information	Strategic communication linked to changes in knowledge, attitudes and skills of the intended audience Seen as part of the social sciences Use multiple channels and products Focused on use (outcomes) rather than production and reach More co-creation
When	At end of the research process	Continuous process where communications is seen as a process for deriving shared meaning, putting due emphasis on regular learning and sharing

Source: Victor and Baca (2011).

In the Nile BDC, partners in Ethiopia used participatory video to allow farmers who could not participate in meetings to have their say in implementing sustainable land management. Local farmers (women) produced the videos to voice their perceptions and opinions on unrestricted grazing, water stress and government-led soil and water conservation. The video received a highly positive response from the local government authority (Cullen, 2011).

Participatory video can help farmers communicate with decision-makers, but alone it is not enough. Higher-level people and local community members need to change their attitudes and be willing to listen to each other (see the summary of the Colombia *conversatorios* in Chapter 6). This requires flexibility and openness on both sides and trust, which is difficult to accomplish (Cullen, 2013).

Another form of communication was adapting scientific models for local users who did not have access to the information. In the Nile BDC, KM specialists and modelers matched land- and water-use practices to landscape-specific needs. They developed a game, modeled on a popular children's card game, which allowed local people to understand and discuss each others' viewpoints on rainwater management more openly (Pfeifer et al., 2012). In the Mekong, researchers used companion-modeling approaches for the same purposes to foster evidence-based dialogue and negotiation by helping different groups who share a water resource to understand each others' viewpoints (Ruankaew et al., 2010).

A number of different communication processes and products were designed for use by high-level decision-makers. In the Mekong, a *State of Knowledge* series of publications was initiated to provide decision-makers with short summaries of the key issues related to hydropower. They included: sediment flows, impacts of hydropower on fisheries, corporate social responsibility and China's influence on hydropower development in the Lower Mekong Basin.

In the Ganges and Nile Basins, donor partners, government decision-makers, and NGOs interested in CPWF research participated in reflection meetings. The meetings tested whether messages met the participants' needs or, if not, how they could be presented more effectively. This is very different from developing messages at the end of the research and then transmitting them to different actors.

We make the following conclusions on the role of communications and KM:

- Because R4D advances through continuous learning, continuous communication and KM are needed to capture and harness what is being learned;
- Because R4D takes place in complex adaptive systems, communications and KM need to deal with ambiguity, learn from failure, value multiple sources of knowledge and help accelerate feedback loops;
- M&E must be designed to capture learning as well as to check for compliance;

- Because R4D emphasizes the use of research outputs or information in engagement with decision-makers, communication and KM are integral parts of research;
- Because engagement is a dynamic and on-going social process, communication and KM must evolve to meet changing needs; and
- Teams must integrate research, communication and KM.

Learning systems

One of the hypotheses that underpins R4D is that learning is a central component. In conventional research, researchers work in controlled settings. In R4D, researchers work on complex problems by engaging with people who may be farmers, investors, government officials or NGO workers. It also implies working at various levels and dealing with wicked problems where there is no one single solution.

The CPWF approach to R4D originated with a set of learning-orientated tools for monitoring and evaluation (impact pathways, outcome logic models, most significant change). This was supplemented with various interactive learning exercises and with knowledge management. Taken together these form a learning system. It is useful to describe the elements of this system and the way different tools influenced learning and how they were interrelated. There is no silver bullet approach for organizing such research. The CPWF's learning system was made up of three broad areas: monitoring, spaces and activities for reflection (opportunities to reflect on progress), and knowledge sharing (Figure 3.3). As Hall (2013) states:

> This however leaves open the question of how one organizes these different activities, organizations and processes in such a way that research plays a valuable role in development. Surely it [cannot] be the same in different countries or subsectors or under different stages of social and market development? The answer is we [do not] know how to organize this, at least not in a specific sense, and this has to be worked out and learned on a case by case basis. The implication of this is that the R4D must have a way of framing this learning.

A program evaluation was carried out in 2013 to learn how these different systems were perceived (Schuetz, 2013). Some of the key findings from the survey were:

- The impact pathways and monitoring and evaluation packages were useful as tools for reflection. But the framework in which they were implemented was quite rigid so that some of them were not carried out.
- Spaces for reflection were important to the learning process that allowed the CPWF to adapt. Innovation funds provided seed money for targeted

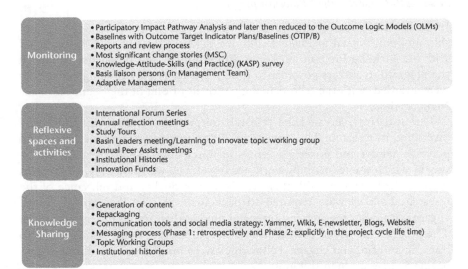

Figure 3.3 The three broad areas of the CPWF's learning system.

Source: Authors.

activities by partners. The reporting and documenting mechanisms were good tools to stimulate reflection.

- Respondents were more positive in face-to-face interviews on topics related to near-term project activities. This was consistent with some projects resisting cross-project integration.
- Knowledge-sharing tools were used less than expected and many of them were thought cumbersome and untested.
- Responses were less favorable to questions on theoretical or less comfortable topics. People were uncomfortable thinking about important long-term strategies to integrate outputs across projects or to translate outputs to outcomes.
- Some Phase 2 concepts were difficult to implement because of the abbreviated time frame. It is hard to focus on learning when there is little time to apply what was learned.

Capacity building

The CPWF used a number of approaches in capacity building. In addition to Masters and PhD students, an innovative approach involved changing the attitudes and behaviors of researchers to implement R4D. This concerned a range of learning and investment in building capacity and competencies across the CPWF and its partners.

A successful model—the fellowship program

In the Mekong, the main thrust of the CPWF's training was through a fellow-ship program started in Phase 1. Funded by AusAID, the CPWF, partnered with the Mekong Program on Water, Environment and Resilience, launched the program to address governance issues. A total of 60 year-long fellowships were awarded, with the fellows being nationals of one of the countries of the Mekong Basin.

The initiative increased the research outputs of the Mekong BDC, and diversified its research. This was important to encourage regional scientists to perceive gender and poverty as issues in water governance. In addition, the way in which the fellowships work was novel. Fellows were assigned mentors who worked together with them to assure the quality and relevance of the research. Fellows were required to participate in CPWF regional events, engage with other research institutes and present their results to potential users. The program provided each fellow with a support system within which to learn, investigate and engage. Fellowships were implemented without fellows having to give up their day jobs

The program helped CPWF to expand its networks, with 60 new contacts working in regional universities, government agencies and NGOs. Fellowship research contributed to on-going CPWF initiatives. Moreover, Fellows are unlikely to leave the region as they often do with training programs leading to higher degrees.

Changing attitudes and behavior of researchers toward R4D

A major effort within CPWF was made to introduce the R4D tools to help research move research outputs to development outcomes.

The first was to introduce tools related to ToC such as PIPA into CPWF research. Lead researchers used a number of tools to analyze what types of impacts their research was supposed to have and who they had to target in order to make this happen (Box 3.8). Many researchers had never used PIPA and these tools. Through network analysis exercises, one senior researcher realized that they were teaming up with the wrong partners if they wanted to achieve impact. Many researchers realized that if their research was to have impact, they must include stakeholders from the outset.

The second approach focused on training and mentoring young researchers. In Phase 2, we replaced the term "student" with "young professional." The change reflected the understanding that young professionals often challenge the established order and encourage the new thinking needed in R4D. In some cases, after graduation they took up key positions within national systems, which extended the influence of the CPWF's R4D.

Box 3.8 Tools used in the PIPA process

- Problem/opportunity tree.
- Visioning.
- Time lining.
- Project outputs.
- Linking outputs to the vision and change that needs to happen.
- Social-network analysis now and in the future.

Conclusions

The CPWF R4D approach evolved over ten years to understand how research could more effectively lead to development outcomes. The experiment was notable for its scope of work (six river basins), extent of partnerships and its independence to explore new modes of working. As Hall (2013) states, CPWF represented an important institutional innovation in the way international agricultural research is used as a tool in the development process.

The CPWF R4D approach explained in this chapter has three distinguishing characteristics that set it apart from conventional research approaches. These characteristics were suited to tackling wicked problems that were identified as the BDCs at different scales within the given basin program.

First was the highly decentralized nature of the basin programs. While they were guided by some over-arching principles described in this chapter, they were allowed to explore and identify entry points as they went along. This was based on the understanding that change is an interactive process, which cannot be planned or controlled as events play out within the basin.

Second was that if research led to development outcomes it had to engage with the development process in a given area. The engagement required investments in time, partnerships and understanding the development environment, and meant reorienting research to be more demand driven, opportunistic and strategic. It meant bringing stakeholders into the research process from the outset so that they helped design the research and felt ownership of the results. It also meant aligning with organizations that had "convening power" or could reach policy-level discussions.

The final characteristic was the need to develop a learning system to enable those involved in the research process the ability to reflect and plan to do better science and development. Here, KM systems such as communication, information management and M&E played key roles in helping different actors within the research frame make decisions based on the best available information.

Acknowledgments

We acknowledge the CPWF basins leaders Miguel Saravia, Pamela George, Kim Geheb, Amy Sullivan, Allen Duncan, Tilahun Amede, Simon Langan, and Olufunke Cofie. Their insights and experience are the basis for this chapter. In addition, we would like to thank CPWF Project Leaders who also contributed insights through their project reports. We would like to thank Ilse Pukinskis, Olufunke Cofie, and Kim Geheb who participated in initial discussions on this chapter.

Notes

1   From en.wikipedia.org/wiki/Emergence (accessed 1 December 2013).
2   Defined as "construction (as of a sculpture or a structure of ideas) achieved by using whatever comes to hand" (Merriam-Webster, 1989).
3   BDC can either refer to the problem set that the challenge represents or to the basin R4D program aimed to address the challenge.
4   Phase 2 projects finish from late 2013 to mid-2014, many in December 2013. Because Phase 2 has been brief, there are as yet few publications that document its progress. We therefore had to rely on unpublished project-progress reports.

References

Alvarez, S., Douthwaite, B., Thiele, G., Mackay, R., Córdoba, D. and Tehelen, K. (2010) 'Participatory impact pathways analysis: A practical method for project planning and evaluation', *Development in Practice*, vol. 20, no. 8, pp. 946–958.

Axelrod, R. M. and Cohen, M. D. (2000) *Harnessing complexity*, Basic Books, New York, NY.

Byerlee, D., Harrington, L. and Winkelmann, D. L. (1982) 'Farming systems research: Issues in research strategy and technology design', *American Journal of Agricultural Economics* vol. 64, no. 5, pp. 897–904.

Candelo, C., Cantillo, L., Gonzalez, J., Roldan, A. and Johnson, N. (2008) 'Empowering communities to co-manage natural resources: Impacts of the *Conversatorio de Acción Ciudadana*', *Proceedings of the CGIAR Challenge Program on Water and Food 2nd International Forum on Water and Food, Addis Ababa, Ethiopia, 10–14 November 2008*, CGIAR Challenge Program on Water and Food, Colombo, Sri Lanka.

Carlile, L., Ballantyne, P., Ensor, J., Foerch, W., Garside, B., Harvey, B., Patterson, Z., Thornton, P. and Woodend, J. (2013) *Climate change and social learning (CCSL): Supporting local decision making for climate change, agriculture and food security*, CCSL Learning Brief No. 1, CGIAR Research Program on Climate Change, Agriculture and Food Security (CCAFS), Copenhagen, Denmark.

Chechi, P. (2012) *Project Volta 4: Governance of rainwater and small reservoirs*, CPWF Six-Monthly Project Reports for April 2012–September 2012, CGIAR Challenge Program on Water and Food, Colombo, Sri Lanka.

Clayton, T. (2013) Final report and evaluation on the documentary film, *Mekong Dam development*, submitted to Sida, unpublished.

Cofie, O. (2012) *Project Volta 5: Coordination and change*, CPWF Six-Monthly Project Reports for April 2012–September 2012, CGIAR Challenge Program for Water and Food, Colombo, Sri Lanka.

Conway, G. (1986) *Agroecosystem analysis for research and development*, Winrock International Institute for Agricultural Development, Bangkok.

Córdoba, D., and White, W. (2011) *Citizen participation in managing water: Do Conversatorios generate collective action?*, CPWF Impact Assessment Series 06, CGIAR Challenge Program on Water and Food, Colombo, Sri Lanka.

CPWF (2010) *CPWF M&E Guide: Theory of change*, monitoring.cpwf.info/background/theory-of-change (accessed May 2013)

CPWF (2011) *CPWF Information and communication strategy*, CPWF, unpublished.

CPWF (2012) *CPWF Significant change stories*, Six-Monthly Project Report for January–September 2012, CGIAR Challenge Program on Water and Food, Colombo, Sri Lanka.

CPWF (2013) *CPWF Institutional histories workshop*, Penang, Malaysia, November 27–29 2012, CGIAR Challenge Program on Water and Food, Colombo, Sri Lanka.

Cross, R. and Parker, A. (2004) *The hidden power of social networks: Understanding how work really gets done in organizations*, Harvard Business Press.

Cullen, B. (2011) 'What is participatory video?', *Proceedings of the CGIAR Challenge Program on Water and Food 3rd International Forum on Water and Food, Tshwane, South Africa, 14 –17 November 2011*, CGIAR Challenge Program on Water and Food, Colombo, Sri Lanka.

Cullen, B. (2013) *Communal grazing land management in Fogera: Lessons from innovation platform action research*, Nile Basin Development Challenge—Rainwater Management for Resilient Livelihoods, nilebdc.org/2013/01/18/ip-fogera-lessons/ (accessed 11 April 2014).

Douthwaite, B. (2002) *Enabling innovation: A practical guide to understanding and fostering technological change*, Zed Books, London.

Douthwaite, B. (2011) 'Harnessing complexity to trigger a blue revolution: the CPWF's theory of change', *Proceedings of the CGIAR Challenge Program on Water and Food 3rd International Forum on Water and Food, Tshwane, South Africa, 14 –17 November 2011*, CGIAR Challenge Program on Water and Food, Colombo, Sri Lanka, cgspace.cgiar.org/bitstream/handle/10568/10387/LiSe005_ns.pdf (accessed 12 April 2014).

Douthwaite, B., Alvarez, S., Cook, S., Davies, R., George, P., Howell, J., Mackay, R. and Rubiano, J. (2007) 'Participatory impact pathways analysis: A practical application of program theory in research for development', *Canadian Journal of Program Evaluation*, vol. 22, no. 2, pp. 127–159.

FAO (2013) *The female face of farming*, Food and Agriculture Organization of the United Nations, fao.org/gender/infographic/en/ (accessed 11 April 2014).

Geheb, K. (2010) *Communication strategy for the CPWF Mekong BDC*, CGIAR Challenge Program on Water and Food, Colombo, Sri Lanka.

Geheb, K. (2011) *Project MK5 Mekong coordination and change project*, CPWF Six-Monthly Project Report for January–June 2011, CGIAR Challenge Program on Water and Food, Colombo, Sri Lanka.

Geheb, K. (2012) *Project Mekong 5: Coordination and change project*, CPWF Six-Monthly Project Reports for April–September 2012, CGIAR Challenge Program on Water and Food, Colombo, Sri Lanka.

George, P. and Meisner, C. (2013), *CPWF BDC progress reports—Ganges*, November 2012–April 2013, CGIAR Challenge Program on Water and Food, Colombo, Sri Lanka.

Gladwin, C. H. (2002) 'Gender and soil fertility in Africa: An introduction, *African Studies Quarterly*, vol. 6, nos. 1 & 2, africa.ufl.edu/asq/v6/v6i1a1.htm (accessed 11 April 2014).

Hall, A. (2013) *The Challenge Program on Water and Food: Opportunities for adding value to experiences using research for development, (R4D)*, CGIAR Challenge Program on Water and Food, Colombo, Sri Lanka.

Hall, A., Dijkman, J. and Sulaiman, R. (2010) *Research into use: Investigating the relationship between agricultural research and innovation*, United Nations University (UNU), Maastricht Economic and Social Research and Training Center on Innovation and Technology (MERIT).

Harrington, L. and Hobbs, P. (2009) 'The rice–wheat consortium and the Asian Development Bank: A history', in J. K. Ladha, Y. Singh, O. Erenstein, and B. Hardy (eds) *Integrated crop and resource management in the rice–wheat system*, IRRI, Los Baños, Philippines.

Harvey, B., Lewin, T. and Fisher, C. (2012) 'Introduction: Is development research communication coming of age?' *IDS Bulletin*, vol. 43, no. 5, pp. 1–8.

Hawkins, R., Heemskerk, W., Booth, R., Daane, J., Maatman, A. and Adekunle, A. A. (2009) 'Integrated agricultural research for development (IAR4D)', *A Concept Paper for the Forum for Agricultural Research in Africa (FARA) Sub-Saharan Africa Challenge Programme (SSA CP)*, FARA, Accra, Ghana.

ICRISAT (International Crops Research Institute for the Semi-Arid Tropics) Eastern and Southern Africa (2011) *2010 highlights*, ICRISAT, Nairobi, Kenya.

Kizito, F. (2012) *Project Volta 4: Sub-basin management and governance of rainwater and small reservoirs*, CPWF Six-Monthly Project Reports for April 2012–September 2012, CGIAR Challenge Program on Water and Food, Colombo, Sri Lanka.

Krebs, V., and Holley, J. (2002) 'Building sustainable communities through network building', Appalachian Center for Economic Networks, orgnet.com/Building Networks.pdf (accessed 11 April 2014).

Lasiter, K. and Stawick, S. (2014) *Linking knowledge: A qualitative analysis of gender and IWRM-related policies in the upper east region of Ghana*, CGIAR Challenge Program on Water and Food, Battaramulla, Sri Lanka, cgspace.cgiar.org/bitstream/handle/10568/35136/Linking%20knowledge%2015%20March%202014%20final.pdf (accessed 24 April 2014).

Lebel, L., Bastakoti, R. C. and Daniel, R. (2010) *Enhancing multi-scale Mekong water governance*, CPWF Project Report Series PN50, CGIAR Challenge Program on Water and Food, Colombo, Sri Lanka.

Leon, C., Douthwaite, B. and Alvarez, S. (2009) *Most significant change stories from the Challenge Program on Water and Food (CPWF)*, CPWF Impact Assessment Series 03, CGIAR Challenge Program on Water and Food, Colombo, Sri Lanka.

Ludi, E., Belay, A., Duncan, A., Snyder, K., Tucker, J., Cullen, B., Belissa, M., Oijira, T., Teferi, A., Nigussie, Z., Deresse, A., Debela, M., Chanie, Y., Lule, D., Samuel, D., Lema, Z., Berhanu, A. and Merrey, D. (2013) *Rhetoric versus realities: A diagnosis of rainwater management development processes in the Blue Nile Basin of Ethiopia*, CPWF Research for Development (R4D) Series 05, CGIAR. Challenge Program on Water and Food, Colombo, Sri Lanka.

Mbabu, A. and Hall, A. (2012) *Capacity building for agricultural research for development: Lessons from practice in Papua New Guinea*, LINK Ltd., Maastricht, The Netherlands.

Merrey, D. J. and Gebreselassie, T. (2011) *Promoting improved rainwater and land management in the Blue Nile (Abay) Basin of Ethiopia*, NBDC Technical Report 1, International Livestock Research Institute, Nairobi, Kenya.

Merrey, D. J. and Cook, S. (2012) 'Fostering institutional creativity at multiple levels: Towards facilitated institutional bricolage', *Water Alternatives*, vol. 5, no. 1, pp. 1–19.

Merrey, D. J., Swaans, K. and Le Borgne, E. (2013) *Lessons from the Nile Basin Development Challenge Program: An institutional history*, CPWF Research for Development (R4D) Series 7, CGIAR Challenge Program on Water and Food, Colombo, Sri Lanka.

Merriam-Webster (1989) *Webster's Ninth New Collegiate Dictionary*, Merriam-Webster, Springfield MA.

Mulligan, M. (2013) 'Modelling a more equitable water future in the Andes', waterandfood.org/2013/10/04/modelling-a-more-equitable-water-future-in-the-andes/ (accessed 11 April 2014).

Olsen, R. (2009) *Don't be such a scientist, talking substance in an age of style*, Island Press, Washington, DC.

PANOS (2007) *Case for communication in sustainable development*, PANOS, London.

Perrin, B. (2002) 'How to—and how not to—evaluate innovation', *Evaluation*, vol. 8, no. 1, pp. 13–28.

Pfeifer, C., Notenbaert, A. and Ballantyne, P. (2012) *The 'happy strategies' game: Matching land and water interventions with community and landscape needs*, NBDC Technical Report 4, International Livestock Research Institute, Nairobi, Kenya.

Quintero, M. (2012) *Project Andes 2: Assessing and anticipating the consequences of introducing benefit sharing mechanisms (BSMs)*, CPWF Six-Monthly Project Reports for April 2012–September 2012, CGIAR Challenge Program on Water and Food, Colombo, Sri Lanka.

Rittel, H. and Webber, M. (1973) 'Dilemmas in a general theory of planning', *Policy Sciences*, vol. 4, pp. 155–169.

Rogers, P. J., Petrosino, A. J., Hacsi, T. and Huebner, T. A. (eds) (2000) *Program theory evaluation: Challenges and opportunities*, Jossey-Bass, San Francisco CA.

Ruankaew, N., Page, C. Le, Dumrongrojwatthana, P., Barnaud, C., Gajaseni, N., van Paassen, A. and Trebuil, G. (2010) 'Companion modeling for integrated renewable resource management: A new collaborative approach to create common values for sustainable development', *International Journal of Sustainable Development and World Ecology*, vol. 17, no. 1, pp. 15–23.

Saravia, M. (2012) *Project Andes 4: Andes coordination project*, CPWF Six-Monthly Project Reports for April 2012–September 2012, CGIAR Challenge Program on Water and Food, Colombo, Sri Lanka.

Sayer, J. and Campbell, B. (2003) *The science of sustainable development: Local livelihoods and the global environment*, Cambridge University Press.

Schuetz, T. (2013) *CPWF learning systems efficiency survey results*, draft, CGIAR Challenge Program on Water and Food, Colombo, Sri Lanka.

Sullivan, A. (2012) *Project Limpopo 5: Coordination and change*, CPWF Six-Monthly Project Reports for April–September 2012, CGIAR Challenge Program on Water and Food, Colombo, Sri Lanka.

Tripp, R., Anandajayasekeram, P., Byerlee, D. and Harrington, L. (1990) 'Farming systems research revisited', in C. Eicher and J. Staatz (ed.) *Agricultural development in the Third World*, The Johns Hopkins University Press, Baltimore.

Ugbe, U. P. (2010) *What does innovation smell like? A conceptual framework for analysing and evaluating DFID-RIU experiments in brokering agricultural innovation and development*, RIU 2010 Discussion Paper 10, r4d.dfid.gov.uk/PDF/Outputs/ResearchIntoUse/riuinnovationsmellutiang.pdf (accessed 24 April 2014).

van Koppen, B., Smits, S., Moriarty, P., de Vries, F. P., Mikhail, M. and Boelee, E. (2009) *The multiple-use water services (MUS) project*, CPWF Project Report Series PN28, CGIAR Challenge Program on Water and Food, Colombo, Sri Lanka.

van Rooyen, A. and Homann-Kee Tui, S. (2009) 'Promoting goat markets and technology development in semi-arid Zimbabwe for food security and income growth', *Tropical and Subtropical Agroecosystems*, vol. 11, pp. 1–5.

Victor, M. and Baca, J. (2011) 'Global information and communication strategic framework', Presentation made at the CPWF Knowledge Management Workshop, 10–13 May 2013, slideshare.net/CPWF/cpwf-communication-and-information-strategic-framework?from_search=4 (accessed 11 April 2014).

Watts, D. J. and Strogatz, S. H. (1998) 'Collective dynamics of "small-world" networks', *Nature*, vol. 393, pp. 440–442.

Weiss, C. H. (1995) 'Nothing as practical as a good theory: Exploring theory-based evaluation for comprehensive community initiatives for children and families', in J. P. Connell, A. C. Kubisch, L. B. Schorr and C. H. Weiss (eds) *New Approaches to Evaluating Community Initiatives*, Aspen Institute, Washington DC.

Whyte, W. F., Greenwood, D. J. and Lazes, P. (1989) 'Participatory action research: Through practice to science in social research', *American Behavioral Scientist*, vol. 32, no. 5, pp. 513–551.

WLE (2013) *Joint mission of the Directors of the CGIAR Research Program on Water, Land and Ecosystem (WLE) and the Challenge Program on Water and Food (CPWF) to Mekong focal Basin*, Vientiane, 25–28 August 2013, Andrew Noble, WLE, and Alain Vidal, CPWF.

4 The institutional history of the CGIAR Challenge Program on Water and Food

Ilse Pukinskis

CGIAR Challenge Program on Water and Food CPWF, Vientiane, Lao PDR; Corresponding author, ipukinsk@gmail.com.

Introduction

From its inception, the Challenge Program on Water and Food (CPWF) was meant to do things differently. In 2002, the CGIAR (formerly the Consultative Group for International Agricultural Research) created three Global Challenge Programs (GCPs) to respond directly to pressing global development concerns. The GCPs were envisioned as pilot programs for the "reinvention of the business model of the CGIAR" (CPWF Consortium, 2002, p. vii). They were to be characterized by their focus on specific outputs, reliance on new partnerships and an inclusive approach to priority setting (CDMT, 2001, p. 6). The CPWF was born into this environment of anticipated change and learning.

In the decade since it started, the CPWF has evolved a set of research-for-development (R4D) approaches. These are processes for undertaking agricultural research aimed at achieving tangible development outcomes (see Chapter 3 for processes and Chapter 8 for examples of outcomes).

Evolution of R4D in the CPWF is best understood through its institutional history. An institutional history is a narrative of how new ways of institutional working evolved to achieve goals better (Prasad et al., 2006). Institutional histories support learning by making knowledge explicit and examining the institutional context within which change occurred. The CPWF's institutional history tells of successes and failures of institutions and individuals.

What can we learn from the CPWF's new way of working to achieve outcomes? In this chapter I analyze how interactions among different players influenced the CPWF's ability to achieve its goals. Successful R4D requires specific capacities at three levels: individual, organizational and institutional (Hawkins et al., 2009). Through good practices at these three levels, R4D can add value to existing research and development processes. Interaction among these levels determined the institutional capacity of the CPWF and shaped its trajectory. The lessons and conclusions I draw from the CPWF's story provide institutional insights for future R4D work.

In the rest of the chapter, I discuss the CPWF's Phase 1, its focus on expanded partnerships and its diverse research on water and food. We go on to describe the role of the Basin Focal Projects (BFPs) in revisiting the CPWF's assumptions. I then show how: (1) Phase 1 and BFP research results; (2) recommendations for focus and coherence by an external review and the CGIAR Science Council; and (3) participatory analysis of impact pathways, stimulated the design of Phase 2 with its emphasis on R4D. Finally, I discuss the challenges that the CPWF faced, its achievements and what I learned about R4D.

Origins of Phase 1

In 2001, a proposal for a Challenge Program on Water and Food was submitted to the CGIAR interim Science Council (iSC) by a consortium of partners with the International Water Management Institute (IWMI) as lead Center. The proposal envisioned "an ambitious research, extension and capacity-building program" with an anticipated 10–15-year timeline. The CPWF's stated development objective was to "increase the productivity of water for food and livelihoods, in a manner that is environmentally sustainable and socially acceptable" (CPWF Consortium, 2002, p. 4). The immediate objectives of the Program were:

1 Food security for all at the household level.
2 Poverty alleviation, through increased sustainable livelihoods in rural and peri-urban areas.
3 Improved health through better nutrition, lower agriculture-related pollution and reduced water-related diseases.
4 Environmental security through improved water quality as well as the maintenance of water-related ecosystem services, including biodiversity.

The CPWF proposed to address water scarcity and related development constraints by increasing agricultural water productivity, that is, producing more food with less water (CPWF, 2005, p. 1). The proposal's business model had five key elements:

1 Consortium Steering Committee: sharing decision-making on strategic management and quality control through the CPWF Consortium of CGIAR Centers, national agricultural research and extension systems (NARES), advanced research institutes and NGO partners;
2 Thematic groups: setting research agendas through communities of practice (thematic groups) in five key and interlinked research themes, coordinated by CGIAR Centers;
3 Benchmark basins: providing geographical focus with an emphasis on regional and local priorities and emphasis on impacts through benchmark basins coordinated primarily by NARES partners;
4 Competitive grants: driving the research agenda forward through competitive grants made from core funds of the CPWF, with grant awards

based on independent peer-review mechanisms to determine merit and alignment with thematic and basin priorities; and

5   Global change agenda: linking to and building on the water-related global change research agenda, primarily through advanced research institute partners (CPWF Consortium, 2002, p. vii).

Phase 1 explored a range of methods to define and solve problems of water and food. It cast its net widely, hoping to identify new means to achieve its objectives. Lessons from Phase 1 were to serve as the basis for the next phases of the CPWF (CPWF Consortium, 2002, p. v).

The CPWF's new approaches for organizing and managing research included "a new quality of partnership" (CPWF Consortium, 2002, p. vi). It posited that collaboration among diverse partners would "lead to break-throughs in how knowledge [solves] problems at basin and field levels." Phase 1 was therefore designed to encourage partnerships beyond the normal CGIAR Center networks. Each of the CPWF's 18 partners, including five CGIAR Centers, was a voting member of the steering committee that took strategic decisions. Three-quarters of the CPWF funding was distributed through competitive grants, which encouraged a broad partnership base. A minimum of one-third of funds for each project was earmarked for NARES partners (CPWF Consortium, 2002, p. vi).

In November, 2002 the CGIAR Executive Council approved the first, 5-year phase of the CPWF, which ran from 2004 to 2008.[1]

Phase 1

Over 200 research and development institutions with natural and social scientists, development specialists and river basin communities participated in Phase 1. Research was focused on five themes in nine benchmark river basins[2] in Africa, Asia and Latin America.

Each theme was led by a CGIAR Center. Themes addressed issues affecting water and food at different scales with different perspectives (CPWF, 2005, p. 6). Themes sought to understand how the main drivers that affected water and food security evolved, and how they might be changed. The five themes and their respective lead Centers were:

- Crop water productivity (IRRI);
- Water and people in catchments (CIAT);
- Aquatic ecosystems and fisheries (WorldFish);
- Integrated basin-level water management systems (IWMI); and
- Global and national food and water systems (IFPRI).

The CPWF proposal hypothesized that research on water and food was best conducted in the context of an entire river basin (Harrington et al., 2006).

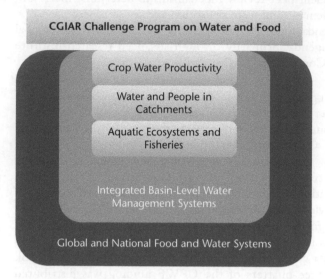

Figure 4.1 CPWF Phase 1 themes at multiple scales.
Source: Author.

The nine benchmark river basins were selected because they represented diverse biophysical, socioeconomic and institutional settings (CPWF, 2005, p. 8). They were: a group of small basins in the Andes (called Andes System of Basins), and the basins of the São Francisco, Volta, Limpopo, Nile, Karkheh, Indus–Ganges, Mekong and Yellow rivers.

With the basin as the main unit, research was to understand the effect of scale at the farm, catchment and basin level on water and food problems. The five themes were interlinked, but were to examine the dynamics of water and food. An integrated, thematic approach to water management was essential to understand how the components of water and food systems interrelate with each other and human activity (Biswas et al., 2007, p. 21). The components include agricultural productivity and sustainability, livelihoods, income distribution and providing ecosystem services. By comparing and contrasting basins, the objective was to draw conclusions at the global level.

Phase 1 contracted research projects to a wide range of institutions. Projects were selected through an independent external review. The first call for concept notes was in March 2003. By October the Consortium Steering Committee (CSC) had approved 50 projects for funding, 31 of which received grants from the CPWF. The remaining approved projects were encouraged to seek additional funding opportunities under the auspices of having already been through the CPWF's rigorous review process. Between 2004 and 2006, the CPWF commissioned two special call projects, ten BFPs, 14 small grants projects (SGPs) and 11 second call projects.

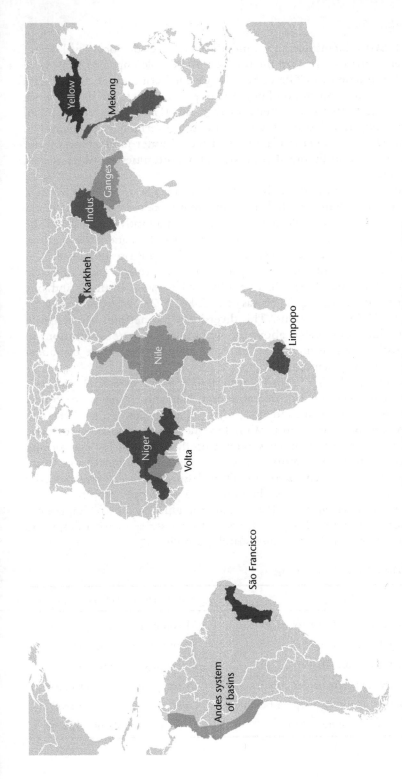

Figure 4.2 CPWF Phase 1 benchmark river basins, plus the Niger Basin, which was included in the BFPs.

Source: CPWF.

The 14 SGPs identified and funded small-scale or local-level water management initiatives that had the potential to be scaled up. The BFPs were established in reaction to CGIAR Science Council criticism of the CPWF's fragmented research portfolio. They documented knowledge in five work packages in the CPWF's nine benchmark basins (Harrington et al., 2006, p. 9) plus the Niger, included at the request of the government of France. The work packages were: water availability and access; water productivity; water poverty; policy and institutional context; and opportunities for water-related interventions.

Over 60 percent of the research funding in Phase 1 was disbursed through competitive grants, which united many different stakeholders (Biswas et al., 2007, p. 2). Projects were selected based on their innovative quality and the diversity and quality of the project teams rather than by national or regional priorities. A total of 68 projects were contracted during Phase 1 (Table 4.1). A median of seven institutions participated in each project, including two CGIAR Centers, four NARES, one advanced research institute and one national or international NGO (Biswas et al., 2007, p. 22). Many projects were led by non-CGIAR institutions. This diversity of partnerships was achieved through specific requirements about the number and types of institutions in each proposal, as well as limitations on the maximum percentage of budget allocated to CGIAR Centers.

The goal of the CPWF's Phase 1 was to support a diverse research program, but the Program received much criticism for its "unfocused" approach. Project selection criteria had generated the project diversity needed to deal with complex water and food issues. With few mechanisms available to foster coherence across projects within basins or across themes, cross-project learning was limited. Basin Coordinators and Theme Leaders struggled to generate coherence. Projects implemented in multiple basins, however, were generally more successful in achieving cross-basin learning *within* their project.

As Phase 1 progressed, the CPWF's understanding of how to implement outcome-oriented research evolved, assisted by a series of workshops and events that it convened to assess its research portfolio.

Table 4.1 The CPWF Phase 1 projects.

Date of call for project funding	Number and type of projects funded
First call (December 2002–October 2003)	31 competitive grant projects
Special call (July–September 2004)	2 commissioned projects
BFPs (first set) (March 2005)	4 commissioned projects
Small grants (August–November 2005)	14 competitive grant projects
BFPs (second set) (February–September 2006)	6 competitive grant projects
Second call (May–December 2006)	11 competitive grant projects

Source: Author.

The CPWF supported innovative research that created new partnerships and mutual learning amongst individuals and organizations. In November 2004 Project Leaders met for their first workshop in South Africa. During the workshop, the CPWF presented an initiative called "knowledge sharing in research." This was a precursor to the participatory impact pathways assessment workshops that would be held in 2006. The workshop was the CPWF's first experiment in running a forum that was not dominated by science.

In November 2006 the CPWF held the first International Forum on Water and Food in Vientiane, Lao PDR. The Forum brought together the CPWF's partners to discuss research results in a format that emphasized synthesis of research experiences and learning. Opinions were mixed; many researchers were dissatisfied with the lack of focus on research and its results. Others approved of the Forum's agenda, which was oriented towards development and outcomes. Disagreement over the balance between research and development focus remained an issue throughout the course of the CPWF. The Forum encouraged participants to overcome their business as usual mindset and to think and debate at individual and program levels.

The CPWF introduced SGPs in an effort to encourage its project partners to begin to think about development outcomes. Run on a short time frame, successful SGPs would demonstrate the types of development outcomes that might be expected from the CPWF's research projects (Woolley, 2011). A call was made to national NGOs and NARES in 2005 to submit proposals for SGPs. The CPWF received 120 proposals, of which 14 were selected (Biswas et al., 2007, p. 23).

SGPs represented only 1.5 percent of the Phase 1 research budget and ran for only 12–18 months. Nevertheless, they "made significant contributions to identifying water and food technology for specific end users (thus showing the potential of the CPWF research in general); to better understanding of adoption; to stimulating research by [NGOs]; and to better connecting the CPWF's researchers in general to the reality of the development process" (Woolley, 2011). The CPWF's flexibility in trying out new approaches for development projects gave it important lessons and insights.

In June 2005 the CSC discussed the gap between impacts predicted in project proposals and those detailed in project reports. It recommended that the CPWF carry out an ex ante impact assessment. The CPWF launched a project "Impact Assessment of Research in the CPWF" in October 2005. The project developed three tools for ex ante evaluation:

- Impact pathways—causal pathways connecting intended project outcomes and impacts with project activities—were incorporated into planning of new projects of the CPWF;
- extrapolation domain analysis identified areas where the CPWF's projects could be scaled up; and
- scenario analysis analyzed possible future events by considering alternative outcomes in relevant conditions at a basin scale (Biswas et al., 2007, p. 53).

Participatory impact pathway analysis (PIPA) workshops held in each basin in 2006 developed systems for monitoring and reporting project outcomes and impacts (PIPA methods are described in Chapter 3). Project teams from second call projects practiced network mapping, visioning and writing log frames oriented toward end users. Visioning exercises allowed participants to see how a project's anticipated research outputs might change the knowledge, attitudes or skills of its target end users. Participants were also introduced to the idea of "most significant change stories": the most notable result they saw in their project. The PIPA workshops were followed by basin-scaling workshops where projects wrote their own most significant change stories.

Through these activities, components of what would become the CPWF's R4D approach began to coalesce. Researchers were asked to think how they expected research outputs to create development outcomes. Impact pathways gave them the tools to do so. Yet although many participants were interested in components of this new approach, uptake was negligible. Work plans and budgets were already set, and projects argued that they had no flexibility in time or personnel to implement this new vision. For many Project Leaders in Phase 1, this was a request by the CPWF that was outside the original call and their planning. Nevertheless, the workshops were an important milestone for the CPWF and were an opportunity to work differently with individual scientists.

The original CPWF management team (MT) was unwieldy with 16 members: Program Coordinator,[3] Program Manager, five Theme Leaders and nine Basin Coordinators. In the early stages of implementation, however, the large MT promoted a program-cohesive team and transparency in procedure development and decision-making. At its fifth meeting in March 2005 the CSC accepted the recommendation of the MT to reduce the MT to the Program Coordinator and Program Manager plus four part-time members, including one representative each of the Theme Leaders and the Basin Coordinators, and two external members (Biswas et al., 2007, pp. 80–81).

The CSC was responsible for the CPWF's governance during Phase 1. It consisted of a delegate from each of the 18 Consortium partners. The 2007 External Review found the CSC setup to be problematic:

> From a management perspective, CSC decisions were perceived to be mainly driven by institutional interests of the [CGIAR Centers] in the Consortium instead of by program interests alone. Some CSC members clearly indicated that their CSC participation was driven mainly by the economic interests of their home institution. Since more than 50% of overall [program] funds remain with the Consortium members, a considerable potential for perceived or real conflict of interest exists. The presence of (economical) institutional interest in CSC decision making has the potential to block critical reform. It was also felt that this setup limited full partner and stakeholder representation in the Consortium.
>
> (Biswas et al., 2007, pp. 3–4)

In February 2008 the CSC assigned CPWF governance to a board, which was established in June 2008 and charged with setting the CPWF's strategic direction and goals. The Board consisted of five members independent of the CPWF Consortium and four representatives of the CPWF's partners, including the Director General of IWMI (P. George, personal correspondence, 27 June 2008). The composition of the Board was designed to overcome the conflicts of interest inherent in the CSC (CPWF, 2009a). The Board made CPWF decisions, but was legally responsible to IWMI, which was otherwise not involved in CPWF governance (P. George, personal correspondence, 27 June, 2008).

The seeds of tension of Challenge Programs within the CGIAR system were sown from the start. Challenge Programs were not independent fiduciary entities, but operated under the organizational umbrella of their respective host Centers. In the case of the CPWF, IWMI was its host Center and served as its legal representative, managed its finances, hosted the secretariat and oversaw overall program management (Biswas et al., 2007, p. 3). The CPWF was embedded within IWMI's institutional structure.

The CPWF's staff members were subject to the administrative policies of their host and partner institutions. While the Program Coordinator (who led the CPWF) reported to the CSC, and later the CPWF Board, it was the Director General of IWMI (later assisted by the CPWF Board Chair) who evaluated their performance. Similarly, some Basin Leaders and the CPWF MT members were employed and evaluated by their respective consortium institutions; the Program Coordinator had only limited authority over them (Biswas et al., 2007, p. 4).

In retrospect, some tension was inevitable. The CPWF was designed as a reform program with an innovative governance and business model (new quality of partnerships; greater partner diversity; steering committee not dominated by Centers; heavy reliance on competitive grants; large funds allocation to NARES). As structured, however, the host Center retained legal responsibility for the hosted program. The CPWF sought flexibility in implementing its plans while IWMI understandably sought to maintain close supervision over CPWF activities. As a reform program, the CPWF was expected to demonstrate the benefits of a "new quality of partnerships" with, however, no guarantee of success. The unstated assumption was that IWMI would be willing to relinquish authority over the CPWF while retaining responsibility for its actions. In the end, this arrangement was difficult to maintain.

Some CGIAR Centers viewed the CPWF's mandate for competitive grants and broader partnerships as an unnecessary burden. These Centers were satisfied with their pre-existing partnerships and resented the time it took scientists to write proposals for competitive funding. Moreover, despite efforts to move beyond the traditional CGIAR partnerships, for projects in many Centers, it was business as usual. The 2007 External Review found projects so strongly linked to their parent Center that they were indistinguishable from

Center-based projects. "That is not to question these projects' merits but rather to question the impact of CP funding as opposed to the operation of the CGIAR Centers in a 'business as usual' setting" (Biswas et al., 2007, p. 41).

Many in the Centers believed that the Challenge Programs competed with them for funding. The CPWF secured new funding of almost US$70 million for Phase 1, but Centers believed they were losing out. The belief was so strong that a 2004 study by the CGIAR Secretariat and Science Council investigated the claim. The study concluded that the Challenge Programs had generated new funding from both existing and new sources that most likely would not have been raised in the absence of the Challenge Programs. The study also noted that funding to Centers in 2003 and 2004 had not declined as a result of the Challenge Programs being established (CGIAR, 2004, p. 15).

In early 2009, the CPWF conducted an online survey to which 76 Project Leaders and staff responded. Most feedback was positive, but respondents identified several weak aspects in Phase 1. Some thought that too many meetings were uncoordinated, and neither time nor money had been budgeted for them. Others thought that initial planning was optimistic, leading to shortages of both time and money as projects wound down.

Many respondents agreed that the CPWF research model worked well and over three-quarters said that the CPWF provided useful training. Eighty-four percent said that in CPWF projects they worked with more and different partners, which three-quarters thought contributed to different science and outcomes. Three-quarters of respondents felt that they achieved different research results, outcomes and impacts than they would have done in business-as-usual research (Sullivan and Alvarez, 2009). This is contrary to the External Review criticism that many CPWF projects were indistinguishable from Center-led projects. The CPWF's approach was resource-intensive and demanding but, according to its researchers, it worked.

Origins of Phase 2

In Phase 1, the CPWF identified options to produce more food with less water through innovations that emerged from its emphasis on diverse partnerships. Through these partnerships, the CPWF redefined how to do effective agricultural research in the face of institutional challenges. The CPWF continued to refine how to operate an R4D program as it prepared for Phase 2.

The results of the BFPs that were released in 2007 called into question the idea that water scarcity was the defining crisis of the new century. They confirmed the importance of increasing water productivity, but emphasized that it was more complex than "more crop per drop." BFP research also showed that the links between water scarcity and poverty were more subtle and complex than previously thought. The BFP results reinforced the reasons for the CPWF to increase its focus on R4D as it headed into Phase 2 (see Chapter 2 for more detail).

As it prepared for Phase 2, the External Review in August 2007 prompted reflection on the CPWF's future direction. The Review found that the nine basins of Phase 1 were too diverse in terms of scale and transboundary politics because the selection criteria were too broad (Biswas et al., 2007, p. 26). The Review also found that most projects lacked a cohesive vision for the basins in which they operated, making basin coordination difficult (Biswas et al., 2007, p. 37). It recommended a re-evaluation of the best way to achieve impact within basins (Biswas et al., 2007, pp. 37–38). On the programmatic level, the Review found that the CPWF lacked "a realistic . . . understanding of its potential impacts," because the original objectives were "visionary rather than [those] against which [program] success can be measured" (Biswas et al., 2007, pp. 54–55). It urged the CPWF to revisit its vision and mission statements.

Overall, the Review was positive and it praised the CPWF's ambitious partnership approach. The Review gave the CPWF's non-traditional approach to research credibility within the CGIAR. In September 2007 the CGIAR Science Council released the External Review and its commentary on it. The Council agreed that the strengthened linkages among CGIAR Centers, NARES, advanced research institutes and NGOs were the most important "added value" of the CPWF (Science Council of the CGIAR, 2009, pp. 2–3). It agreed with the Review's suggestion that the CPWF should re-evaluate its objectives and develop a more cohesive approach to research in river basins. The Science Council (SC) recommended approval for Phase 2:

> In sum, subject to the development of a well-conceived and more tightly focused strategy and implementation and monitoring plan for Phase 2, as well as a clear exit strategy and timeline, the SC endorses continuation of the CPWF. As a next step the SC looks forward to reviewing and endorsing a Phase 2 plan at the SC '09'.
> (Science Council of the CGIAR, 2009, p. 4)

Priorities for Phase 2 were set based on lessons learned from Phase 1, the External Review's recommendations, the Science Council commentary and consultation with colleagues at IWMI. The groundwork for Phase 2 planning took place prior to the release of the External Review report at a meeting in January 2007 with the CSC, MT and Basin Coordinators. A second planning meeting in October 2007 with a more limited attendance consolidated the CPWF's plans for Phase 2. Participants agreed that it was necessary to reduce the number of basins and refocus the work within some of them. They also agreed that the CPWF needed to invest more heavily in communications, which was weak in Phase 1. They noted the need to distinguish between different audiences and to build relationships with policymakers. They emphasized the need to use impact pathways and network maps to identify key stakeholders to ensure the CPWF's success (CPWF, 2007, pp. 1–8).

Some Phase 1 projects were important in the design of Phase 2. With no explicit R4D framework they had produced research outputs that could translate to outcomes. Their project teams understood the development side of their research and the process to produce outcomes. Many of these projects were led by researchers who were long-term champions of specific topics (Woolley and Douthwaite, 2011).

By early 2008, the CPWF had defined a new structure for Phase 2. Its objective was, "To increase the productivity of water for food and livelihoods in a manner that is environmentally sustainable and socially acceptable" (CPWF, 2007, pp. 1–8). To overcome the lack of cohesion amongst projects that was criticized in Phase 1, research in Phase 2 would focus on water-related development challenges in six river basins (CPWF, 2008, p. 4).

The six Basin Development Challenges (BDCs) were founded on successful Phase 1 projects that had high potential for impact. Moreover, their focus was the nexus between poverty and water (CPWF, 2010, p. 17). Priorities within each BDC were set drawing partly on Phase 1 research. The CPWF also carried out a specific process of consultations to identify the most pressing challenges in each of the six basins (CPWF, 2009c, p. 1).

The CPWF consultation process to identify the BDCs lasted from November 2007 to June 2009. Basin Coordinators identified two priority research areas for each of their basins. These were evaluated against priorities identified through parallel consultations with stakeholders in each of the basins. Selection of the BDCs was heavily influenced by the technical results of Phase 1 research. Feedback from the External Review and the recently published *Water for food, water for life: A comprehensive assessment of water management in agriculture* (Molden, 2007) also featured in BDC design. Selection criteria included stakeholder agreement on the challenge's importance, their motivation to work on it, the CPWF's ability to contribute and high-impact potential (CPWF, 2010, p. 20). Draft BDCs were submitted to the CPWF MT for comment and approval. The final BDCs are listed below (Table 4.2).

Inception workshops were held in each basin. During the workshops, stakeholders identified the research questions that needed to be addressed to tackle a particular BDC. These questions were then divided amongst four or five projects that fitted together as an integrated basin program. Each BDC was led and maintained by a coordination and change project, which supported the Basin Leader. Basin Leaders fostered change, built networks and adjusted project objectives based on emerging opportunities and learning (CPWF, 2010, p. 21). All BDC programs were required to show that their organizational plans matched their expected outcomes. They were also required to incorporate learning mechanisms that allowed them to react to emerging opportunities (CPWF, 2009c, p. 2).

Selection of Basin Leaders was difficult because of the demanding nature of their role. Basin Leaders were charged with providing leadership in project design and implementation related to generating and evaluating outcomes and impacts (CPWF, 2009d, p. 10). The CPWF MT debated the ideal skill set for Basin Leaders. These individuals had to be good networkers, have a command of technical knowledge related to the particular BDC and possess leadership

Table 4.2 The CPWF BDCs.

River Basin	Basin Development Challenge
Andes System of Basins	To increase water productivity and reduce water-related conflict through the establishment of equitable benefit-sharing mechanisms
Ganges	To reduce poverty and strengthen livelihood resilience through improved water governance and management in coastal areas of the Ganges Basin
Limpopo	To improve smallholder productivity and livelihoods and reduce livelihood risk through integrated water resource management
Mekong	To reduce poverty and foster development by optimizing the use of water in reservoirs
Nile	To strengthen rural livelihoods and their resilience through a landscape approach to rainwater management
Volta	To strengthen integrated management of rainwater and small reservoirs so that they can be used equitably and for multiple purposes

Source: Author.

qualities. Their ability to guide their research program toward a cohesive solution to their BDC was key to the CPWF's ability to achieve outcomes.

The CPWF added Topic Working Groups (TWGs) to Phase 2. TWGs were communities of practice to facilitate cross-basin learning and train basin teams through sharing experiences and mentoring (CPWF, 2011a). As world leaders in their fields, TWG leaders would feed cutting-edge thinking and methods into the TWG communities. Four TWGs were selected: resilience; global drivers of change; learning to innovate; and modeling and spatial analysis. The CPWF's researchers were encouraged to form TWGs around shared interests as Phase 2 research developed, strengthening the quality of research on that particular topic. Membership in TWGs would emerge from project team members working on elements of the topic focus (CPWF, 2009d, p. 4).

Phase 2

Phase 2 of the CPWF started in November 2008 as the second International Forum on Water and Food in Addis Ababa, Ethiopia marked the end of Phase 1. With the inception of Phase 2, a new set of challenges to the operation of a coherent R4D program became evident.

The CPWF adapted its management and governance structures for Phase 2. In keeping with its decentralized approach, it promoted a horizontal structure. The CPWF program team was made up of a number of interlinked teams: the MT, administration and finance, knowledge management and research management. The CPWF filled the newly created positions of Research Director and Innovation and Impacts Director. At the basin level, Basin Leaders coordinated and facilitated integration and learning amongst individual projects. Project Leaders implemented their respective projects but were encouraged to work as integrated teams in their BDC.

CPWF governance was comprised of two main bodies: the newly established CPWF Board and the CSC. The Board's role was to provide oversight and strategic vision for the Program. The CSC retained limited functions including selection of, and providing strategic advice to, the Board; responsibility for high-level consultation; and control over the Joint Venture Agreement (CPWF, 2010, pp. 31–32).

The first three basins to plan and contract their projects were the Nile, the Mekong and the Andes. Most projects funded under these BDCs were selected through a competitive process similar to Phase 1. Proposals for each project were evaluated by at least three independent reviewers, whose recommendations guided the selection (CPWF, 2009b, p. 9).

During project implementation in 2009 it became clear that achieving coherence and collaboration across projects within a basin could not be assumed to occur despite BDC design and the contracting process. The CPWF had envisioned that the activities and outputs of all projects within a basin would complement and reinforce each other as a collective effort. Based on a

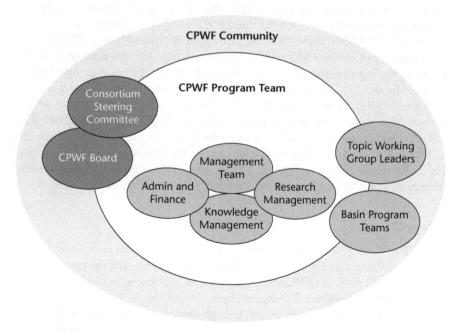

Figure 4.3 Governance and management structures of the CPWF in Phase 2.
Source: Author.

common vision of the BDC and how it could be addressed, projects were designed to perform complementary functions and foster interdependence. Learning was intended to be across projects in a basin, as well as across basins.

Many projects focused on their own activities and compliance milestones, however. They gave little attention to opportunities available through mutual support, sharing and learning. In some cases, they thought that cross-project collaboration was unworkable and mechanistic, with apprehension replacing complementarity and common vision.

Cross-project collaboration was constrained by concerns about possible problems of timing and synchrony. Researchers feared that progress of their project might be hindered if it relied on inputs from other projects. There were examples of projects ignoring relevant outputs already produced by sister projects. Connections among projects were idiosyncratic and often only emerged when a project recognized the need for information. Throughout Phase 2, there was less collaboration and information sharing across projects in basins where projects were selected through competitive bidding.

When contracting the second round of basin projects (Volta, Limpopo and Ganges) a few months later, the CPWF opted to commission the majority of projects. It argued that commissioning projects ensured competent organizations and people were involved, and built on the social capital of the CPWF's Phase 1 "community of practice." It was also necessary given time and budgetary constraints. The CPWF evaluated potential institutions against criteria that included the institution's record in leading similar initiatives and its governance and financial management (CPWF, 2010, p. 21). It was a new approach to "leveling the playing field" amongst stakeholders, recognizing relationships between partners as important in creating holistic basin programs.

In the Volta, Limpopo and Ganges basins, the CPWF attempted to avoid the disconnect among projects that had resulted from the process of selection in the first round of BDC contracts. It used writeshops for proposal development to create cohesive BDC research programs in the Volta, Limpopo and Ganges. During the writeshops projects were expected to agree on a common vision, select common research sites and confirm expectations for collaboration. It was only then that project proposals were sent for external review. Even then, cross-project collaboration continued to be a challenge. In some cases different site-selection criteria meant projects were unable to agree on common sites.

The Ganges was the last BDC commissioned. An external team of experts helped develop terms of reference for institutions identified as candidates for commissioning (Ruvicyn et al., 2011). The benefits of including cross-project complementarities into the Ganges BDC design were evident early on. Successful BDC programs had projects that shared a common vision and coordinated as they worked toward it.

CPWF management recognized that an integrated research, innovation and impact strategy needed effective knowledge management. Knowledge management has to manage research outputs to influence stakeholder attitudes, skills and behavior, and in doing so produce outcomes. The CPWF's knowledge management team worked with the BDCs to ensure that they incorporated impact pathways and theories of change. Communications, monitoring and evaluation (M&E) and information management formed the basis for the CPWF's knowledge management work.

The Phase 2 knowledge management framework emphasized communications, aiming to communicate better and publicize the programs. Communication strategies introduced at the program level proved useful in some basins, and less useful in others. Due to the decentralized nature of the CPWF, each basin program developed its own communication plan and strategy, which resulted in varying levels of success across basins. Basin-level communication activities did achieve impact when they were able to translate science results into the "languages" of different target stakeholders.

In February 2010 the CGIAR announced its plans to create a series of CGIAR research programs (CRPs). The CRPs were intended to align the research of the 15 CGIAR Centers and their partners into efficient, multidisciplinary programs. In April 2010 the CGIAR Consortium Board (CB) recommended that the CPWF be integrated into the newly formed, IWMI-led CRP on Water, Land and Ecosystems (WLE). It was agreed that the CPWF would continue to operate independently through the end of 2013, in the process contributing to the WLE program. In August 2011 the CPWF Consortium Steering Committee was dissolved and the boards of the CPWF and IWMI merged (CPWF, 2011b, p. 1). This was a major turning point for the CPWF from both a programmatic and governance perspective, and reduced the CPWF's programmatic independence.

WLE took a thematic approach to program design and centered the program on five thematic strategic research portfolios (SRPs) operating in ten river basins (including the six BDC basins) (WLE, 2011). R4D with an outcome orientation was featured in the WLE strategy for partnerships and capacity building.

WLE was launched in February 2012. Over the next year and a half, strategies for the closure or continuation of BDC activities post-CPWF varied a great deal. In some basins, WLE is likely to provide continuity for impact-generating research on the CPWF basin challenges while in other basins regional networks have offered to provide continuity.

The CPWF's Phase 2 was designed to move from outputs to outcomes, followed by Phase 3, which was intended to move from outcomes to impact as the Program's exit strategy. The CGIAR CB decided to integrate this exit strategy into WLE. This decision was presented as a fait accompli in a meeting of the CB with Center Directors General and Board Chairs. In taking this decision, the CB appears to have judged that integration with WLE would yield a similar result as the CPWF's planned exit strategy. The CPWF has continued to work with WLE to build on the CPWF's successes where appropriate so as to keep generating outcomes and impact.

With plans for a Phase 3 (2014–2018) canceled, the CPWF found itself in an awkward situation. It had entered into Phase 2 recognizing that R4D and innovation systems take more time than allowed by short-term projects. Although the CPWF had generated development outcomes, as discussed in other chapters of this book, it had counted on a Phase 3 to move further from outcomes to impacts. Integrating Phase 3 into WLE raised the prospect of possible disruption in the trajectory of R4D impact in some basins. Just when it was coming to understand the complexities and nuances of the approaches it was advocating, the CPWF was terminated. Jointly with WLE, the CPWF redoubled its efforts to ensure that its most promising outcomes became key elements of WLE's portfolio.

Adaptive management (discussed in Chapter 3) was a principle of the Phase 2 strategy, but the CPWF's capacity to change course in this case was limited. The most important limiting factor was that the time allocated for Phase 2 was too brief to allow for the full cycle of adaptive management.

Many project staff saw the CPWF's compliance systems as burdensome, a criticism that plagued it throughout its existence. While Project Leaders thought reporting requirements were too elaborate and took too much time, their reports often did not provide adequate information to the CPWF MT. Adaptive management was more successful, however, within basin programs where flexibility allowed projects to capitalize on learning.

Budget cuts imposed by CGIAR in early 2012 were a particularly difficult challenge for the CPWF. The CPWF was required to reduce its 2012 budget by 37 percent. It did so by reducing program spending by 45 percent and basin spending by 21 percent. The 45 percent cut in program spending closed the TWGs, just as they were beginning their activities. As a result, the CPWF's global learning and synthesis strategy was hampered by its inability to create a mechanism for cross-basin reflection. The CPWF was not able to engage in certain outcome-oriented processes, particularly organizational learning.

The worst effect of the budget cuts, however, was the damage to the CPWF's reputation with basin partners. The CPWF's intention was always to build a program based on a "new quality of partnerships" (CPWF Consortium, 2002 p. vi). It was accepted that the CPWF's success would depend on its ability to foster relationships and trust. The CPWF built its social capital by engaging with stakeholders and partners over ten years. In all active basins, on-going projects now had to reduce their budgets, which caused frustration amongst project partners. These large and unexpected cuts threatened to erode the social capital that the CPWF had accumulated with partners. The CPWF Program Director visited all basins and met with almost all project teams in 2012 in an effort to mitigate the damage resulting from the budget cuts. The repercussions of the budget cuts, however, remained tangible through the end of the CPWF and gave a lesson: where it is difficult to build trust, rebuilding it is a lot more challenging.

In Phase 2, the CPWF set out to test its version of R4D through a unique program design. Its decentralized, basin-centered framework was an attempt to depart from the traditional organizational structures of the CGIAR. Yet in the end, the CPWF's own hierarchical management structure showed that it failed to move beyond the classic CGIAR model. A strong central team was needed to get BDCs established. But as the BDCs became operational, many functions of the MT became redundant. The MT never transformed from coordinating program setup to coordinating program-wide learning and leading (CPWF, 2013). Due to the creation of the CRPs, the dissolution of the CPWF board and the 2012 budget cuts, the MT operated continuously in crisis-management mode.

At the CPWF's final peer assist[4] in June 2013, several Basin Leaders discussed the impact of the strategic decisions that the MT made in an attempt to keep the CPWF operational. Many felt that the dissolution of the CPWF board and budget cuts had a systemic impact on relationships within the CPWF (CPWF, 2013). Many projects were reluctant to bridge the void left by an otherwise occupied MT. In many basins, projects were unwilling to work beyond the confines of their required milestone and compliance deliverables. This resulted in a failure to forge a common, cross-project vision of how to address the identified BDCs.

WLE focused much of its efforts in 2013 on developing its focal region approach. By the end of the year, a focal region strategy emerged that built upon the CPWF process and structures. The CPWF's basin coordination teams from the Nile, Ganges, Volta and Mekong were asked to lead the co-ordination of the four focal regions prioritized by WLE (East Africa, South Asia, West Africa and Southeast Asia). A letter circulated by the Director General of IWMI following an IWMI board meeting in December 2013 officially noted the importance of building upon CPWF emerging outcomes and structures.

The CPWF concluded on 31 December 2013 but project and basin activities and learning continue in other forms. Local, national and international partners, more often than not, have positioned themselves to carry forward the best of the CPWF's activities and achievements.

Conclusions

Despite setbacks and challenges, the CPWF made substantial progress in generating a richness of research outputs. It transformed these outputs into development outcomes, defined as modifications in decision-maker knowledge, attitudes and skills resulting in changes in policy or practice. Other chapters of this book expand upon that progress. But the value of an institutional history lies in making tacit knowledge explicit and examining the institutional context within which change has, or has not, occurred. What then can be learned from this story?

For more than a decade the CPWF tested an R4D program in an institutional environment subject to the forces of the systems, organizations and individuals in which it operated. In this respect, the CPWF was not unique. It was not the first agricultural R4D program that claimed to "do something differently," nor will it be the last. The institutional history of the

CPWF is a story of institutional change and the lessons it provides offer insight for future R4D efforts. Some of the key lessons follow.

The strengthened capacity of individuals who took part in CPWF is a key outcome of the CPWF, and is an important legacy

The success of an R4D project is as much about strengthening capacities to respond to the needs of stakeholders as it is about water productivity, food production or improved livelihoods. There were various wins in influencing scientific thought in the direction of development outcomes over the course of the CPWF. The multi-disciplinary nature of the CPWF's research raised researchers' awareness of the value of working in diverse teams, and sometimes they were more willing to do so. Sometimes the CPWF succeeded in breaking down the barriers and silos that surround the agricultural and water disciplines, but in other instances it failed to do so. Yet the CPWF's legacy is a vision of what successful R4D looks like. That vision was adopted by many who worked with the CPWF, and they will be the ones to carry it forward in their future work.

Institutional challenges to pursuing research for development initiatives within the CGIAR are not new. As discussed in Chapter 7, the CGIAR has a history of initiating institutional reform, beginning in the 1970s when it recognized the need to tackle commodity-based farming systems. The creation of the GCPs was the third round of CGIAR reform, and the CRPs in 2010 is the fourth[5] (CGIAR, 2011). We suggest that the proponents of reform consider this repeating story.

R4D approaches and lessons must be communicated in languages that are logical and relatable to their intended audiences

The CPWF spent more than a decade learning about the science of water, agriculture and poverty as well as the science of R4D and innovation. Some of the things it did worked better than others, but it learned useful lessons. Throughout this time the CPWF sought to articulate these lessons in such a way as to sway skeptics. During the transition to the WLE CRP, the CPWF did not effectively communicate the virtues of its approach—nor its difficulties. Now, however, WLE is confronted with the same challenge as that faced by the CPWF 10 years ago: how to translate research outputs into development outcomes.

Science can, and must, inform and complement development processes

If scientific research is to be useful, it must produce information (outputs) that can be used to influence people to change what they do or how they do it (generate outcomes). In Phase 2, it became clear that those BDC programs that became conveners were most successful in achieving outcomes. Convening BDCs were able to use research outputs in engagement processes to change knowledge, attitudes and skills, that is, outcomes as defined by the CPWF. It takes time and energy to develop the social capital needed to be able to bring different people together and be a conduit for research outputs. Science can inform engagement but science without engagement represents lost opportunities to achieve impact and outcomes.

Acknowledgments

The full institutional history of the CPWF is as diverse and unique as each of the individuals that have contributed to it. I thank the following people for taking the time to share their CPWF experiences and memories with me: Jennie Barron, Boru Douthwaite, Kim Geheb, Pamela George, Amanda Harding, Larry W. Harrington, Tonya Schuetz, Alain Vidal and Jonathan Woolley. I thank Sharon Perera for her invaluable help in tracking down a decade's worth of program documentation. Finally, I thank Andy Hall and Michael Victor for their contributions in shaping the framework and style of this chapter.

Notes

1   2003 was designated as an inception year.
2   Increased to ten when the Niger River basin was included in the BFPs.
3   The title of "Program Coordinator" was changed to "Program Director" in late 2007, although the functions remained the same.
4   Starting in 2011, annual three-day peer assists brought together Basin Leaders and program-level staff to discuss issues identified from the desk review of project reports.
5   The rounds of reform were: (1) establishment of NRM within Centers, (2) Ecoregional and Systemwide Programs, (3) Challenge Programs and (4) CRPs.

References

Biswas, A. K., Palenberg, M., and Bennet, J. (2007) *External Review of Challenge Programme on Water and Food*, CGIAR, Washington DC, cgiar.org/exco/exco13/exco13_cpwf_cper.pdf (accessed 3 December, 2013).

CDMT (Change Design and Management Team) (2001) *Designing and managing change in the CGIAR: Report to the mid-term meeting 2001*, worldbank.org/html/cgiar/publications/mtm01/mtm0105.pdf (accessed 8 April 2013).

CGIAR (2004) *Synthesis of lessons learned from initial implementation of the CGIAR pilot Challenge Programs: A joint report by the Science Council and the CGIAR Secretariat*, Independent Science & Partnership Council (ISPC) Secretariat, Rome, sciencecouncil.cgiar.org/fileadmin/templates/ispc/documents/Publications/2e-Publications_Reviews_ChallengePrograms/SC_CP_LessonsLearned-PilotProcess_Oct2004.pdf (accessed 22 August 2013).

CGIAR (2011) *The CGIAR at 40: Institutional evolution of the world's premier agricultural research network*, CGIAR Fund Office, Washington DC, library.cgiar.org/bitstream/handle/10947/2549/cgiar@40_final_LOWRES.pdf (accessed 9 April 2013).

CPWF (2005) *CGIAR Challenge Program on Water and Food: Research strategy 2005–2008*. CPWF, Colombo, Sri Lanka, cgspace.cgiar.org/bitstream/handle/10568/5402/ CPWF%20Research%20Strategy%202005-2008.pdf (accessed 9 April 2013).

CPWF (2007) 'Notes from the Stockholm Workshop on planning for CPWF Phase 2, October 4–5, 2007'. Internal Document, CPWF, Colombo, Sri Lanka.

CPWF (2008) *CPWF research agenda and implementation plan for Phase 2: An update for the Science Council, 12 March 2008*, CPWF, Colombo, Sri Lanka, cgiar.org/ www-archive/www.cgiar.org/exco/exco14/exco14_cpwf_phase2.pdf (accessed 11 June 2013).

CPWF (2009a) 'CPWF Board meeting minutes'. Internal Document, CPWF, Colombo, Sri Lanka.

CPWF (2009b) *Contracting the CPWF Basin Development Challenges*, CPWF, Colombo, Sri Lanka, infoandina.org/sites/default/files/news/files/FINAL_GENERAL_ GUIDELINES.pdf (accessed 14 April 2014).

CPWF (2009c) *Defining the CPWF Basin Development Challenges (BDCs)*, CPWF, Colombo, Sri Lanka.

CPWF (2009d) *Medium term plan 2010–2012*, CPWF, Colombo, Sri Lanka, cgspace. cgiar.org/bitstream/handle/10568/4708/CPWF_MTP_2010-12.pdf (accessed 14 April 2014).

CPWF (2010) *2009 CPWF annual report*, CPWF, Colombo, Sri Lanka, cgspace.cgiar. org/bitstream/handle/10568/5409/CPWF_Annual_Report_2009.pdf (accessed 9 April 2013).

CPWF (2011a) *Adapting to change, changing how we do research*, CPWF, Colombo, Sri Lanka, cgspace.cgiar.org/bitstream/handle/10568/12575/CPWFbooklet_nov11. pdf?sequence=1 (accessed 14 May 2013).

CPWF (2011b) 'CPWF governance and management options with CRP5', Internal Document, CPWF, Colombo, Sri Lanka.

CPWF (2013) '2013 Peer assist notes', Internal Document, CPWF, Colombo, Sri Lanka.

CPWF Consortium (2002) *CGIAR Challenge Program on Water and Food full proposal*, r4d.dfid.gov.uk/Output/182299/Default.aspx (accessed 14 March 2013).

Harrington, L. W., Gichuki, F., Huber-Lee, A., Humphreys, E., Johnson, N., Nguyen-Khoa, S., Ringler, C., Geheb, K. and Woolley, J. (2006) *Synthesis 2006*. CPWF, Colombo, Sri Lanka, cgspace.cgiar.org/bitstream/handle/10568/16708/CPWF_ Synthesis_2006.pdf (accessed 14 May 2013).

Hawkins, R., Heemskerk, W., Booth, R., Daane, J., Maatman, A. and Adekunle, A. A. (2009) *Integrated agricultural research for development (IAR4D)*, kit.nl/net/KIT_ Publicaties_output/ShowFile2.aspx?e=1626 (accessed 19 February 2013).

Molden, D. (ed.) (2007) *Water for food, water for life: A comprehensive assessment of water management in agriculture*, Earthscan, London and International Water Management Institute, Colombo, Sri Lanka.

Prasad, C. S., Hall, A., and Thummuru, L. (2006) *Engaging scientists through institutional histories,* Institutional Learning and Change (ILAC) Institute, Brief 14, cgiar-ilac. org/files/publications/briefs/ILAC_Brief14_institutional.pdf (accessed 11 September, 2013).

Ruvicyn, B., van Brakel, M., Douthwaite, B. and Harrington, L. (2011) *CPWF proposal development workshop: Ganges Basin*, CPWF, Colombo, Sri Lanka.

Science Council of the CGIAR (2007). *Commentary on the first External Review of the Challenge Program on Water and Food (CPWF)*, CGIAR Science Council, Rome, r4d.dfid.gov.uk/PDF/Outputs/WaterfoodCP/CPWF_ER_SC_Commentary_21_09_07.pdf (accessed 14 April 2014).

Sullivan, A. and Alvarez, S. (2009) *Learning from Phase I: A survey of project leaders and staff*, CPWF, Colombo, Sri Lanka, cgspace.cgiar.org/bitstream/handle/10568/5399/CPWF_Learning_from_Phase1.pdf (accessed 9 April 2013).

WLE (2011) *CGIAR Research Program 5: Water, Land and Ecosystems*, CGIAR, Montpellier, France, iwmi.cgiar.org/CRP5/PDF/Water_Land_Ecosystems/CRP5_Water_Land_and_Ecosystems_20110926.pdf (accessed 3 December, 2013).

Woolley, J. (2011) *Review of the CPWF small grants initiative*, CPWF, Colombo, Sri Lanka, cgspace.cgiar.org/bitstream/handle/10568/4198/IA09_reviewsg_sept_web.pdf (accessed 11 June 2013).

Woolley, J. and Douthwaite, B. (2011) *Improving the resilience of agricultural systems through research partnership: A review of evidence from CPWF projects*, CPWF, Colombo, Sri Lanka, results.waterandfood.org/bitstream/handle/10568/12495/IA10_sg_wolley Douthwaite_web_dec11.pdf (accessed 15 July 2013).

5 Innovating in a dynamic technical context

Larry W. Harrington[a]* and *Martin van Brakel*[b]

[a]CGIAR Challenge Program on Water and Food CPWF, Ithaca, NY, USA;
[b]CGIAR Research Program on Water, Land and Ecosystems WLE,
Colombo, Sri Lanka; *Corresponding author, lwharrington@gmail.com.

Introduction

In this chapter we discuss Challenge Program on Water and Food (CPWF) research on technologies and how it was used to define and research problems of water scarcity, water productivity, poverty, food security, livelihoods and the environment. We place technological research within the CPWF's theory of change (ToC) and show how we used learning to inform engagement with decision-makers. We place learning in a longer-term, dynamic context taking account of previous and concurrent research.

We then explore how technical innovation and institutional change complement each other. The CPWF's experience is that we need both when the purpose of research is to get to outcomes. We describe how the CPWF explored how people innovate and its effects on productivity, income, gender and equity, resilience and ecosystem services. We show how we used this to inform the processes of engagement and dialogue. We conclude with an analysis of categories of technologies that the CPWF studied. We comment on the research problems of intensifying and diversifying agroecosystems and their relationship to improved rainwater management.

An overview of CPWF project categories

The CPWF planned and implemented 123 projects, 68 in Phase 1 and 55 in Phase 2 (Box 5.1). It selected Phase 1 projects from proposals received in response to a competitive call for research on defined topics. Projects were for the most part independent and were not intended to form coherent programs in basins. Phase 2 projects were designed as integrated basin-level programs to address specific development challenges.

A note on sources

Like the other chapter authors, we faced a dilemma in selecting sources. Research in CPWF Phase 1 projects (2004–2008) was published in books and journals, but did not use a research-for-development (R4D) approach. Research

---

Box 5.1 Breakdown of the CPWF projects by Phase

Phase 1 (2004–2008):
Research projects—first call for proposals for competitive grants—31
Research projects—second call for proposals for competitive grants—11
Basin Focal Projects—10
Small Grant Projects—14
Commissioned projects—2

Phase 2 (2009–2013):
Basin Development Challenges projects for Andes, Ganges, Limpopo, Mekong, Nile and Volta basins—43
Research Into Use projects—4
Innovation fund small grants—8

---

in Phase 2 (2009–2013) did use R4D and we use it to describe translating research outputs into outcomes. But at the time of writing in late 2013, Phase 2 projects were still incomplete. There were fewer refereed papers, so reference to Phase 2 projects comes partly from unpublished sources. We discuss research on technologies from both Phase 1 and Phase 2 using an R4D framework.

Technologies for outcomes

CPWF has addressed issues of water, food and livelihoods in basins by using outputs from research to engage with policymakers. CPWF started by focusing on problems of water scarcity, water governance and water productivity, but later focused on water-related Basin Development Challenges (BDCs). CPWF research on technologies in Phase 2 was in the context of the principles of R4D discussed in Chapter 3 (Box 5.2).

CPWF's research on R4D technologies was oriented towards outcomes. It generated outputs that would be useful in engagement and dialogue with decision-makers; enhance partner and stakeholder knowledge, attitudes and skills; and foster stakeholder decisions to change practice.

Figure 5.1 is a conceptual framework that shows where technical research fits in the larger innovation process. Outputs consist of knowledge that contributes to better understanding a development challenge, and informs designing strategies to address that challenge. Both understanding and strategies aim to support decision-making and negotiation. Outcomes are when people decide to modify what they do or how they do it.

"Define the problem" refers to breaking down the basin development challenge into problems that research can solve. "Learning selection" (discussed

Box 5.2 **The principles of research for development (see Chapter 3)**

• A focus on encouraging people to innovate rather than only on research products and technologies.
• Innovators are embedded and interdependent in the local institutional, policy and political contexts.
• Focuses on a defined objective within a development challenge using an adaptive approach.
• Users are involved so that their knowledge, perspectives and needs are integrated into the project.
• Analysis and taking account of lessons learned.
• Science and research is not at the center of the work, but is part of the social process by which people innovate.

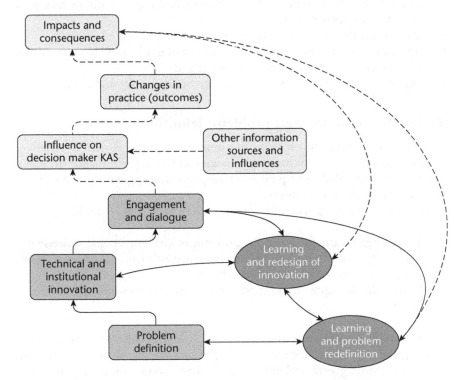

Figure 5.1 Conceptual framework showing where technical research fits in the process of innovation. KAS = knowledge, attitudes and skills. Research outputs are more likely to change outcomes (decisions, practice) when they influence KAS.

Source: Authors.

in Chapter 3) is shown by feedback loops where learning contributes to defining the problem and what innovations are likely to help solve it. "Engagement," "influence on decision-maker KAS" and "changes in practice (outcomes)" are uses for research outputs.

Changes in practice and policies, which are outcomes, occur when decision-makers alter their knowledge, attitudes or skills (KAS). The question is how to use outputs to influence people's KAS and in doing so contribute to outcomes favorable to development. Production of research outputs (shown in dark gray in Figure 5.1) is under the control of a project. The process of translating outputs to outcomes (shown in light gray) is not. Decision-makers confront many influences other than being informed of research results, which makes the process harder.

Figure 5.1 is similar to the boundary framework of Chapter 6, where research outputs are used to create better understanding, and for decision support and negotiation support. Within its ToC, CPWF justifies technical research because:

• research to define problems [development challenges] without trying to solve them is incomplete;
• it is not prudent to try to solve problems that are not understood;
• it is fruitless to do research whose outputs will not be used in engagement [the social process of informing people]; and
• engagement not guided by evidence can be ineffective and inefficient.

Technical innovation and problem definition

The purpose of research on technical innovation is to provide knowledge, including knowledge about new technologies and their performance. In R4D, knowledge is intended to help inform engagement, dialogue and negotiation. It should be credible and relevant.

There are three categories of knowledge from technical research:

1   Knowledge on the nature of the problem or challenge being addressed;
2   Knowledge on the design and performance of technical options (often integrated with institutional options); and
3   Knowledge (ex ante and ex post) on the broader consequences of innovation.

Many CPWF projects invested a lot in defining problems. This involved defining the issues or challenges in question and tracing chains of cause and effect, sometimes giving problem trees. We have known for a long time that understanding the causes of a problem often suggests ways to solve it (Tripp and Woolley, 1989). We give some examples.

Box 5.3 Main findings of the CPWF Basin Focal Projects (Fisher and Cook, 2012)

- Water is less scarce than the capacity to use it.
- Only a small proportion of rainfall goes through agriculture, even in dry basins.
- Water scarcity at the farm level has many dimensions not always linked to average rainfall.
- There is large and inexplicable variability in water productivity, especially in rainfed systems.
- Water productivity is useful as an indicator but is less useful as a goal.
- Relationships between water scarcity and poverty are subtle and complex with many intervening variables.
- Water-related entry points for research are influenced by development trajectories.
- Institutional factors often determine whether available water is or is not used well.

Basin Focal Projects

The Basin Focal Projects (BFPs) were about problem definition. BFPs analyzed, quantified and characterized problems of water availability and scarcity, water productivity, water and poverty, and water and institutions in ten river basins. BFP findings resulted in fundamental changes in understanding within and outside the CPWF regarding the nature of water-related global challenges (Box 5.3).

In the Mekong Basin

The CPWF carried out 19 projects in the Mekong Basin, many aiming to understand better interrelated problems of water, food and hydropower. Project PN50 in the Mekong pinpointed governance problems affecting dam and reservoir design. Because institutions did not talk to each other, when hydropower authorities designed new infrastructure, they often ignored problems that were well known from previous experience (Molle et al., 2009; Lebel et al., 2010).

Project Mekong 3 in Laos identified the possibility of man-made floods along a cascade of reservoirs in a catchment. When the reservoirs are full and there is heavy late-season rainfall, dam operators must communicate to coordinate the release of overcapacity. If they do not, the resulting floods can damage infrastructure and crops and can kill people and livestock (Ward et al., 2012).

In coastal areas of the Mekong, project PN10 defined and characterized conflicts in land and water use between farmers. Some wanted freshwater for rice production while neighboring farmers wanted saline water to produce shrimps (Tuong and Hoanh, 2009). The solution was to zone the land for each use and to operate sluice gates to provide each zone with water of the quality it needs.

In the African basins

Projects in the African basins covered landscape and rainwater management, small reservoirs and large dams, supplemental irrigation and alluvial aquifers, water access and integrated water resource management, and market-value chains for crops and livestock.

Project Nile 2 in the Nile looked into why farmers in Ethiopia did not adopt strategies to conserve soil and water recommended to stop degrading the land. The reason was the difficulties and sensitivities to reconcile community-level, bottom-up needs for land use with top-down priorities and strategies for land management (Ludi et al., 2013). This was consistent with the finding in a parallel project, Nile 1, in the same basin (Merrey and Gebreselassie, 2011). We can understand the problems of conserving soil and water if we know how they are influenced by institutions and policies, which we then must reconcile with community needs. The process also needs adequate institutional, technical and financial capacity at all levels.

Project PN19 in the Nile found that management of large dams often has undesirable social and environmental impacts, including increased incidence of malaria (McCartney, 2009). Further research showed that dam managers could modify operations giving substantial benefits to human health but at negligible cost in terms of lost hydropower generation. Research outputs were not used to engage with dam operators, or energy and health sector policy-makers, however, and were not translated into outcomes.

Project PN28 in the Limpopo and other basins identified the global problem that most water systems are designed for single use by direct consumption or irrigation. These designs, however, reduce water productivity and constrain household incomes. Defining the problem suggested that introducing water systems designed for multiple uses increased water productivity and improved livelihoods (Mikhail and Yoder, 2008; van Koppen et al., 2009a; van Koppen et al., 2009b).

In the Ganges Basin

Projects in coastal areas of the Ganges in Bangladesh give an example of dynamic problem definition. The area consists of polders, which are parcels of land of tens or hundreds of hectares, surrounded by embankments (dykes). Polders are islands surrounded by tidal rivers, protected by the dykes from flooding during the rainy season and from saline water during the dry season.

Farms in polders were hindered by lack of freshwater during the dry season. Many grew only one low-yielding monsoon crop of rice each year, sometimes followed by a low-yielding, non-irrigated legume crop during the dry season.

The project started with the problem defined as "a lack of freshwater to intensify cropping." This evolved to become "a lack of integrated water management within polders to enable intensified cropping." This in turn evolved to "a lack of water infrastructure within polders to allow integrated water management and intensified cropping." The first objective required improved infrastructure to manage water within the polders. Land cropped with high-yielding rice in the wet season must be drained to sow dry-season crops, which must be irrigated with stored freshwater.

Projects PN10 and Ganges 2 identified three intensified cropping possibilities: two crops of high-yielding rice per year; a rice crop followed by high-value dry-season crop; or rice followed by culture of shrimp plus fish in brackish water where the river was salty in the dry season. In some parts, river water is fresh almost year round, which allows triple cropping with rice, or rice rotated with another crop. Success of the intensified system, however, depends on draining off excess water at the end of the rice season, and storing freshwater and then closing the sluice gates before the rivers become too saline (Sharifullah et al., 2008; Humphreys, 2012).

Project Ganges 3 found that overcoming problems of water management depended on new technologies of crop and water management and new institutions. These are needed to coordinate management of sluice gates and within-polder infrastructure design. Moreover, government institutions at several levels have to coordinate and rationalize strategies of polder maintenance and investment (Mukherji, 2012b).

In the Yellow River Basin

Aerobic rice was developed to replace irrigated rice in areas where water had become too scarce. Project PN16 extended the technology to traditional upland cropping areas (maize, soybean) in China where in some years crops are damaged by waterlogging and flooding (Bouman, 2008). Taking account of the spatial and temporal incidence of the problem, the project designed strategies including aerobic rice as well as upland crops to lessen farmers' risk of crop loss.

Project PN42 on groundwater governance in Asia found that informal sharing and thriving groundwater markets were useful for poor smallholders. They improved productivity and high-value intensification in the water-scarce countries of the basins studied.

Who defines the problem?

In research to define problems and development challenges, CPWF used multiple sources of information (Box 5.4) in multiple iterations. Because of

---

Box 5.4 Sources of information used in defining problems in the CPWF

- Information from past research (Merrey and Gebreselassie, 2011);
- results from scenario modeling and spatial analysis;
- peer knowledge and monitoring and evaluation;
- key informant and stakeholder consultation;
- outputs from innovation platforms and learning alliances (see Chapter 3);
- participatory video (Chowdhury et al., 2010) (see Chapter 3); and
- diagnostic tools developed for community use (Sellamuttu et al., 2010; DAE, 2012).

---

the dynamic nature of R4D, each cycle of learning results in improved understanding and re-defining the problem.

Nonetheless, the question of "Who defines the problem" is not easily answered, especially when problems are contentious. Moreover, power balances influence which key informants and stakeholders are consulted and who is influential in innovation platforms and learning alliances. It is easy to introduce biases that confuse important issues related to gender, equity and ecosystem services. Researchers should be aware of the possibility of bias in defining problems. They must be sensitive to which social groups will benefit at the expense of which other groups.

Summary

A scan of the 120 CPWF projects listed in the Appendix shows that nearly all projects had defining the problem as a major item. The depth of analysis used by projects to define problems, however, varied a lot.

The broader dynamics of innovation

Some new projects assume that researchers start with a clean slate and that there is little to learn from antecedent projects. Research managers and funders require projects to demonstrate impact, driving them to assert that their work will find important new knowledge. This knowledge will produce widespread and significant impacts, which will be wholly attributable to the project. The narrative implies that nothing existed before the project and that future success will occur only because of the project. Stated in this form, the narrative is flawed.

Literature on innovation in R4D sees things differently. Innovation is a social process that is risky and unpredictable, where success comes from learning from failure. New ideas are tested, accepted or discarded in a dynamic and iterative process of learning selection (Douthwaite, 2002; Perrin, 2002). This learning process takes more time than is available in a short-duration project. A history of the successful introduction of conservation agriculture in rice-wheat systems in the Indo-Gangetic Plains describes the setbacks and unanticipated directions (Harrington and Hobbs, 2009). The process took more than 20 years.

Learning selection can be applied to complex systems such as agroecosystems (Douthwaite and Gummert, 2010). Learning strategies can be structured into theories of change and formalized into outcome logic models (Alvarez et al., 2010, see Chapter 3). This implies that researchers will be well served if they can foresee the trajectory of innovation using time frames longer than that of their current project. This allows them to build on past learning and to link up with concurrent projects on similar issues being managed by other partners.

The experience of the CPWF is that using systems of innovation in a longer time frame is often the key to progress and ultimate success. We describe three examples from CPWF projects. In all three examples, engagement and the social processes of innovation were as important as the technical research. Institutions and policies were as important as technologies and getting to outcomes covered multiple projects only some of which were managed by the CPWF. Moreover, the innovation trajectory covered 8–10 or more years.

Slash and mulch

Project PN15 in Central America worked on the replacement of slash and burn on hillsides with slash and mulch (locally named Quesungual slash and mulch agroforestry system, QSMAS). Both feature long-term, multi-year rotations, with grain crops alternating with tree regrowth. Slash and burn involves slashing a plot, letting plant residues dry, and then burning them, leaving nutrient-rich ash but little soil cover. QSMAS also involves slashing (and removing valuable wood for other purposes while preserving key tree saplings), letting plant residues dry, and then planting crops directly into the soil cover. The project showed the benefit of slash and mulch on soil fertility, soil microorganisms, soil water-holding capacity, improved drought tolerance, reduced erosion, reduced risk of landslides during heavy rain, more rapid tree re-growth, reduced deforestation, higher crop yields, increased productivity and higher family incomes (Pavon et al., 2006; Castro et al., 2009).

PN15 did not discover or invent QSMAS, which comes from work in the early 1990s when the Food and Agriculture Organization of the United Nations (FAO) and other partners developed it with farmers in the village named Quesungual in Honduras. QSMAS proved itself during the El Niño drought of 1997 and Hurricane Mitch in 1998. There was reduced drought damage and little erosion and no landslides during the hurricane.

Project PN15 first reviewed and synthesized existing information collected by the FAO project 1995–2005. Based on this synthesis, the project carried out research to accomplish three tasks: (1) measure and quantify the ecological and economic consequences of adopting QSMAS; (2) identify technical and socioeconomic factors governing adoption of QSMAS; and (3) foster scaling-out of the practices into other parts of Honduras, Nicaragua and Guatemala. It took its place in the longer-term trajectory of a self-propelled innovation process (CPWF, 2012a). Work on QSMAS continues in other forms and with other sources of funding.

Goats and fodder

In the dry Limpopo Basin, project Limpopo 3 worked on the problem of goat deaths. Goats in Zimbabwe are largely produced by poor households in marginal areas and goat sales are important in their livelihood strategies. In local and national markets, there is unmet demand for goat meat, which in the past has not been reflected in prices received by farmers. Moreover, animal deaths exceed 20 percent per year, often due to lack of fodder in the dry season. Animal quality was low and animals were often sold in distress sales where a farmer will take whatever (low) price the buyer offers (van Rooyen and Homann-Kee Tui, 2008; van Rooyen and Homann-Kee Tui, 2009).

Recent innovations have transformed this picture. Innovation platforms (see Chapter 3) involving farmers, merchants, researchers and other stakeholders promoted formal goat auctions on defined dates. The auctions were held in pens designed for small animals and equipped with scales and small ramps. Buyers competed and paid much more for animals of higher quality, with prices received by farmers at times increased five- or sixfold. This motivated farmers to improve goat management with housing, fencing, investment in improved feed and fodder and in animal health. As auction pens have proliferated, there are fewer goat deaths and fewer distress sales (van Rooyen, 2012).

The irony is that for many years prior to the auction system, researchers had worked on alternative sources of dry-season fodder for goats, but farmers showed little interest. With return to goat farming transformed and the value of dry-season fodder greatly increased, farmers now invest their own resources in dry-season fodder and seek research support.

CPWF only became involved in 2010 with the launch of the Limpopo BDC. But innovation platforms and research on auction pens dated from 2005 through a series of projects funded by Germany, the EU and South African Development Community. CPWF helped researchers work with farmers to identify suitable sources of dry-season fodder that was now in demand. CPWF supported and accelerated progress in an autonomous social process of innovation with its own trajectory.

Permits and pumping

In West Bengal, smallholders have difficulty in accessing shallow groundwater to intensify irrigated cropping. The difficulties were not technical but bureaucratic and legal; farmers needed a government permit to obtain an electrical connection and install a small pump. When they got the permit, they still had to pay the cost of wires, poles and a transformer for their connection, as well as the pump. The permit system was justified as a means of avoiding over exploiting groundwater, but in practice it was fraught with rent-seeking and corruption (Mukherji, 2008).

Researchers in project PN42 identified areas, called safe blocks, where there was no risk of depleted groundwater because of high annual recharge. Researchers then informed policymakers how costly the permit system was. They showed that it was an obstacle to the government priority to foster higher productivity and expand the irrigated area in West Bengal. The result was a policy change in which farmers in 301 safe groundwater blocks no longer needed permits. Their pumps must be less than 5 horsepower and discharge less than 30m³/hour. The state electricity authority made electrical connections less expensive by connecting farmers for a fixed fee of Rupees1000–30,000 (US\$16–475, September 2013) depending on the connected load (PN42 and PN60) (Mukherji et al., 2012).

The result has been rapid growth in the use of small pumps and irrigated farming in West Bengal. The lead researcher, Aditi Mukherji, received the Norman Borlaug Award for Field Research and Application for this work (Mukherji, 2012a).

What was the role of the CPWF? The Government of India for decades had prioritized intensive farming irrigated with groundwater in the eastern Ganges. Research identified safe zones, quantified the cost of misguided policies and used this knowledge to engage with policymakers. The projects built on previous research of PN42 in Phase 1 and PN60 in the BFPs. The research continues in the International Water Management Institute with support from the Gates Foundation. Again, CPWF was one partner among many, but contributed to a process of innovation created by research.

Summary

A scan of the 120 CPWF projects listed in the Appendix shows that more than 30 projects focused on technical innovations and were also designed for longer-term, self-propelled social processes of innovation.

Technologies and institutions

Experience from CPWF projects is that technical innovation and institutional change often go together, and that one without the other does not usually reach outcomes. In discussing the topic, we touch on issues discussed in Chapter 6.

Farms and markets

Even in straightforward projects on new varieties and changed fertilizer use, technical change was linked to institutional change. Project PN2 in Eritrea found that village-based seed enterprises were needed to produce and distribute the project's new varieties (Grando et al., 2010). (We cannot confirm their success and institutional sustainability.) Project PN1 in the Limpopo worked with fertilizer sellers to make fertilizer available in smaller bags, which smallholders could afford (Dimes et al., 2005). Project PN7 in the coastal Ganges, selected salt-tolerant varieties for salt-affected areas. It needed farmers to participate to select varieties and for less stringent rules to release them (Ismail, 2009; CPWF, 2012g). Goat auctions and pens in project Limpopo 3 in Zimbabwe discussed above were an institutional innovation that transformed goat farming and the livelihoods of many poor families.

Community action

There are more examples in which community action was needed to enable technical change. The success of QSMAS discussed above depended on community self-enforcement of an absolute ban on burning residues. This is because one fire can destroy the benefits of QSMAS for the whole community (PN15). Water systems designed for multiple uses are more productive than those designed for single uses as discussed above. Community involvement is necessary, however, for the successful management of multiple use systems (PN28).

Project PN35 in Bangladesh focused on improving community-managed fish culture in areas that flooded in the wet season but were used by many smallholders for cropping in the dry season. Improved technologies, such as stocking with fingerlings, mesh at water exits to reduce escapes, and careful timing of fish harvest to give time for the fish to grow helped increase productivity. Communities introduced practices that gave more equitable access to the fisheries; landless poor have unlimited fishing rights, but only with hook and line, not with nets. The success of these practices, however, was dependent on communities enforcing their own rules, and maintaining lease rights to the seasonally flooded areas (Sheriff et al., 2010; CPWF, 2012e).

See Chapter 6 for more on community action.

Policies

CPWF projects emphasized the importance of links between policy, institutions and technology.

Rapid increase of smallholder crops irrigated by groundwater irrigation in West Bengal, discussed above, came from government policy changes. Informed by research, the government eliminated pumping permits and fostered electricity connections (PN42, PN60). Modeling in project PN10 on land-use zoning reduced conflicts between rice growers and shrimp producers in coastal Vietnam. The research led to changes in government policies to use the land-use zoning identified by the modeling (CPWF, 2012d). Zero-till drills adapted to the conditions of dryland farm systems in the Yellow River in project PN12 were included in a government subsidy program for farm equipment (CIMMYT, 2010). These are among many examples.

An example from the Andes

An example of the tight interrelationships among policies, institutions and technologies comes from the Andes. The Andes basins that flow from east to west are short and steep. There are the usual upstream–downstream externalities that affect different groups with different interests in different ways (Box 5.5).

Everyone could be better off if institutional mechanisms were established to share water-related benefits and costs. Benefit-sharing mechanisms (BSMs) can generate funds from downstream water users to encourage upstream land and water management practices with positive externalities (and discourage practices with negative externalities). BSMs can provide incentives to use improved technical practices such as replacement of intensive hillside tillage with no-till agriculture, introduce tree crops and many others.

In these instances, technical innovation depends on incentives made available through institutional change, for example trust funds for highland investment and incentives. BSMs are pointless if they are not directed towards technologies that improve or maintain water quality and reliable water flow to downstream users. The policy context is also favorable as national policies in many Andean countries give high priority to maintenance of alpine ecosystem services, and to reduce poverty in highland communities.

CPWF, together with the Centro Internacional de Agricultura Tropical (CIAT), the Consortio para del Desarrollo Sonstenible de la Ecoregión Andina (CONDESAN), and other partners, has played key research and engagement roles since 2005 in the development of several BSMs (PN22, Andes 1, Andes 2, Andes 3 and Andes 4) (Estrada et al., 2009; Johnson et al., 2009; Quintero et al., 2009; Escobar and Estrada, 2011; CPWF, 2012f).

Box 5.5 Wants of different groups with different interests in the Andes

- Downstream urban dwellers want clean, reliable water supplies.
- Lowland farmers want cheap, reliable irrigation water of suitable quality.
- Midstream hydropower companies want reliable water with low silt content.
- Sports enterprises want clean reliable year-round water.
- Upstream highland communities want improved livelihoods.
- Civil society in general wants to maintain important highland ecosystem services and environmental flows.

We emphasize a further example. Partnered with the Ministry of Environment of Peru, CPWF projects, in particular Andes 2, defined priority areas and designed a BSM for the Cañete River basin. The Ministry is using the Cañete case study as a pilot project. The Cañete BSM is establishing a trust fund to finance its activities (Quintero, 2012). CIAT and the CPWF partner, CONDESAN, worked with the Ministry and with public and private companies and NGOs. The objective was a BSM that provided equitable benefits from the use of ecosystems and provided direct benefits to rural communities. The findings are being scaled-out to over 30 river basins in Peru through a Remuneration Mechanism for Ecosystem Services hosted by the Ministry with CONDESAN's support.

CIAT and CPWF helped draft national legislation for ecosystem services valuation. It requires valuation of water uses to be included in all environmental impact studies for new public and private development projects.

Summary

A scan of the 120 CPWF projects listed in the Appendix indicates that the majority of CPWF projects discussed the importance of links between technical innovation and institutional change.

Assessing the consequences of innovation

We earlier defined three functions for technical research in the context of R4D. One of these was to generate knowledge (ex ante and ex post) on the broader consequences of new practices. We discuss this further.

An R4D project should anticipate ex ante and assess ex post the consequences that it will bring about. Only then can we tell if potential outcomes are harmful or helpful, for whom, and in what ways. It is important to know what the consequences of innovation are if it is to be used to guide decision-makers, inform negotiators or otherwise in autonomous social processes.

Innovation and change can have many subtle and unexpected consequences. The challenge for research is to choose which of them are important to anticipate *ex ante* and which to assess *ex post*. Many consequences are difficult to measure and are therefore often not monitored. They are better addressed through periodic selective impact assessment/evaluation. A major challenge is the choice of parameters to check and when and how often to review them.

The list of performance indicators (Box 5.6) is intimidating and some variables are more readily measured than others. We are not surprised that no CPWF projects assessed the complete list of variables, although we find some examples for most categories. Box 5.6 lists a few of them.

Human health

Several projects explored the consequences of innovation for human health. In project PN19 in Ethiopia, researchers determined how different scenarios in

---

Box 5.6 **Innovation may have many consequences, including changes in the following:**

- Crop or enterprise yields (land productivity);
- other input productivity (water productivity, labor productivity);
- livelihood strategies, including market interactions and labor migration;
- incomes, poverty and food security;
- farm household profits and their distribution within the household;
- climate and market-related risk;
- human health;
- gender and equity, including access to resources;
- cross-scale consequences and externalities (e.g., undesirable downstream effects of a practice favorable for upstream users;
- water availability for other uses (quantity, timeliness, quality);
- system sustainability and resilience;
- build-up or loss of social capital;
- land and water resource quality and whether it is regenerating or degrading;
- ecosystem services, biodiversity and the environmental quality; and
- economic, social and environmental costs of the innovation at multiple scales.

---

managing large dams affected the incidence of malaria, although the outputs were not translated into outcomes (Kibret et al., 2009; McCartney et al., 2009). In Burkina Faso, project PN46 explored how different ways of managing small reservoirs affected water quality and human health (Boelee et al., 2009; Andreini et al., 2010). The project also developed tools for communities to monitor water quality and health indicators (Andreini et al., 2009). In project PN51 in Ghana researching the use of wastewater in urban and peri-urban irrigation, its impact on human health was a key issue (IWMI, 2009; CPWF, 2012b).

Gender and equity

Projects assessed the consequences of innovation on gender and equity in different contexts.

In South Africa (PN28) and elsewhere, water systems designed to take account of small-scale homestead irrigation and other productive activities can be favorable to women. "Homestead-scale [multiple-use water services (MUS)] not only [meet] domestic water needs but . . . give women a greater say over productive activities at home . . . [H]omestead-scale MUS [are] the best way of using water for productive self-employment that . . . includes the poor and women" (van Koppen et al., 2009b).

In the central Nile Basin, modifying livestock and range management strategies can have large consequences for the whole region. Project PN37 developed an analytical tool that assesses the consequences of change on gender. The tool features the following steps (Van Hoeve and van Koppen, 2005):

1   Analyze the role of each animal in the livestock production system to determine which animals are most valuable for men and for women.
2   Predict what the expected impacts on the gendered costs and benefits will be when a specific technology is introduced in a water scarce area.
3   Use as a tool at different levels (community, development agent, researchers) for communities to analyze the importance and role of livestock in their lives, as it relates to water, to stimulate mutual understanding about the importance and limitations of livestock rearing.

In the same region, innovative strategies for range management—for example, fencing off areas to allow pasture to regenerate—improved incomes for some families. But poor women who rely on pasture commons for their livelihoods, were worse off (Nile 2) (Cullen, 2013).

Based on a scan of CPWF projects, Figure 5.2 summarizes projects emphasizing different categories of consequences.

Categories of technologies studied by the CPWF

The CPWF funded 120 projects over ten years in its two phases. Projects ranged in duration from 1 to 4 years, with budgets varying from US$18,000 to

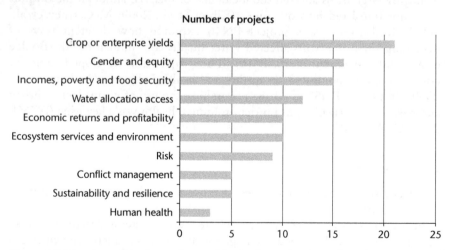

Figure 5.2 Number of projects that emphasized monitoring selected categories of consequences.

Source: Authors.

US$2.5 million. They have published over 1200 papers and reports.[1] There will be more as completed projects continue publishing results and others finish. We classify CPWF research into Phase 1 Projects (2004–2008), BFPs (2005–2009) and Phase 2 projects (2009–2013).

CPWF projects carried out research on a broad range of technical innovations. It was broad because of the numerous high-quality proposals submitted for funding in its two phases. In assessing them, CPWF: interpreted the concepts of water scarcity and water productivity broadly, sought innovative approaches to research on water and food, and included non-traditional research partners (see Chapter 1).

The following sections give examples of CPWF technical innovations. This complements Chapter 6, which focuses on institutional innovation. We describe categories of technical innovation in CPWF projects and give examples which cover farm to basin scales, and include land and water management in irrigated and rainfed systems. They also include crop, livestock and fish management, with attention to water use in intensive and diversified systems. We provide examples of the range of CPWF research on technical innovation by summarizing categories of innovation and their distribution across projects (Figure 5.3). The distribution was based on an analysis of project reports and only includes projects that focused on technical innovation. Most of them fell into several categories. Numbers in parentheses in the subsequent text refer to the number of projects where the particular innovation was the primary focus of the project. They provide a sense of the relative emphasis given by CPWF to different technologies.

Diversifying and intensifying agroecosystems

Diversifying (a wider range of enterprises) or intensifying (more enterprises per year) agroecosystems were common topics. Projects emphasized vegetables or trees (17), livestock (9), and fisheries or aquaculture (9), seven of which were on fishing in tropical reservoirs. The driving forces were not technical changes in land and water management, but opportunities for higher incomes through market-value chains. The strategies to intensify/diversify drove modified land and water management. Research was responsible to examine the consequences of these changes on sustainability, resilience, equity and ecosystem services at different scales.

We discussed project Limpopo 3 to intensify goat production above. It led to demand for technology for improved production of dry-season fodder, which was available, but had never been adopted by farmers. Project Ganges 2 sought to transform the intensity and diversity in coastal polders, described above. Farmers went from one low-yield monsoon rice crop per year to three high-yielding rice crops, or two rice crops and one fish crop per year. The requirement was institutional with technical changes to improve water storage and control. The MUS project (PN28) showed that single-use water systems can be managed to produce high-value vegetables in the off season, which

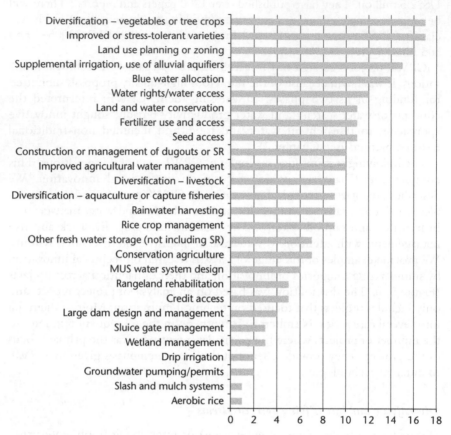

Figure 5.3 The number of CPWF projects focusing on different technical innovations. SR= small reservoirs, MUS= multiple-use water services.

Source: Authors.

benefits women. Changing the policy for pumping permits (PN42) allowed smallholders in West Bengal to use groundwater to irrigate and intensify their cropping systems.

Varieties and crop management

A number of projects introduced improved varieties (17), usually combined with improved crop management (11), fertilizer (11) and credit (5). About half of them were on irrigated rice (9).

Another group of projects focused on rainfed staple crops under abiotic stress. Project PN1 in the Limpopo Basin defined problems in terms of risk (flood and drought), market deficiencies, and lack of draft power to apply low-till or water harvesting. It aimed to address the problems by introducing drought-tolerant crop varieties, access to seed and to fertilizer in small, cheaper packs, and rain-water harvesting. The project made no breakthroughs (Siambi, 2011).

Project PN2 in hilly, degraded land in Eritrea defined problems as drought stress, high fertilizer prices and land degradation. Technical innovations were drought- and disease-tolerant crop varieties; fertilizer use; and practices to conserve land and water. The project invested in participatory plant breeding, variety selection involving women and village-based seed production. The institutional sustainability of the seed systems, and whether farmers adopted the new practices is not clear (Gomez-Macpherson et al., 2009; Grando et al., 2010). Project PN6 in the Volta tested varieties, seed, fertilizer, credit and rainwater harvesting (Asante, 2011).

Low yield and water productivity of rainfed staple crops in drought-prone areas is a difficult problem. Projects tested crop varieties, soil fertility, water harvesting, and input and product markets but struggled to get good results. Water harvesting was laborious and risky, although it did not include small reservoirs, groundwater for irrigation or building structures with machines. These projects evolved from the Centers' (International Crops Research Institute for the Semi-Arid Tropics, International Center for Agricultural Research in Dry Areas and Centro Internacional para Mejoramiento de Maíz y Trigo) on-going work (Low et al., 1991; Kumwenda et al., 1996; Waddington et al., 1998; CIMMYT 1999).

A second group of projects explored how crop varieties and management can deal with special conditions. For example, a project in the Mekong introduced "drawdown" cropping in hydropower reservoirs that are drained seasonally. The cropping calendar depends on the timing of the drawdown, which is predictable, if risky. Project Mekong 1 tested early-maturing varieties of rice and cassava with farmers and is scaling out a new cassava variety in Vietnam (Sellamuttu, 2012).

Aerobic rice is another special condition. Project PN16 in China identified and mapped environments where in some years there is not enough rain for flooded rice and in other years too much for upland crops. Aerobic rice, which can grow well either waterlogged or dryland, is a solution. The project tested varieties of aerobic rice and crop management with the aim to extrapolate the technology to other parts of Asia (Bouman, 2008; CPWF, 2012c).

Salty groundwater near the soil surface in the coastal Ganges is another special condition. Project PN7 used available salt-tolerant varieties of rice, oilseeds, pulses, fodder crops and vegetables to increase yields and intensify and diversify farm systems. Salt-tolerant crops could be irrigated with groundwater in the dry-season farming, which improved food security. Women participated in selecting crop varieties. The intensified system gave more work for women, good for some women and not for others. Over-strict rules on releasing crop varieties was an institutional constraint (Castillo et al., 2007; Islam et al., 2008; Ismail, 2009; CPWF, 2012g).

Improved varieties and better crop management, rotations, drainage and water control transformed productivity and intensity of rice-based agroecosystems in polders in coastal Bangladesh (PN10, G2, G3) (Sharifullah et al., 2008; Tuong and Hoanh, 2009; Humphreys, 2012).

Land and water in landscapes

Another group of projects worked at the whole farm to landscape level. They focused on rainwater harvesting (9) or improved water management (10). Some involved conservation agriculture in rainfed systems (7) including the innovative slash and mulch (QSMAS) to replace slash and burn systems on hillsides. These were complemented by others focused on conserving land and water (11). Also working at a landscape level were projects on land- and water-use zoning (17), rehabilitating rangeland (5) and wetlands management (3). Projects using rainwater harvesting often overlapped with the variety, fertilizer, water, markets group described above.

Water storage, irrigation and drainage

Projects worked on the design and management of small reservoirs (10) or other water storage (7), such as inside polders that are surrounded by saltwater in the dry season. A related topic was sluice-gate management for water control and drainage (3). There were projects on MUSs, which use water better (6) or develop under-utilized groundwater resources (3). There were projects that focused on irrigation methodology including supplementary irrigation and the use of alluvial aquifers (15), two of which studied drip irrigation.

Dams, reservoirs and access

Finally, there were projects that studied reform of policy on water access and how it affected farm-level water use (12), basin-level changes in blue water allocation (14) and large dam design and management (4). Of the latter, some focused on individual reservoirs while others considered cascades of reservoirs in a river basin.

Key lessons

The experience of the CPWF in research on technologies leads us to several conclusions.

In R4D, the role of research on technical innovation is to generate credible and relevant knowledge. Such knowledge is used to inform engagement, dialogue and negotiation and thereby translate research outputs into behavioral outcomes. *Outputs* consist of knowledge that contributes to understanding a development challenge better, and informs strategies designed to address that challenge. Both understanding and strategies aim to support decision-making and negotiation. *Outcomes* are when people decide to modify what they do or how they do it.

Both technical research *and* the use of its outputs in engagement processes were important because:

- Research to define problems (development challenges) without trying to solve them is incomplete.
- It is not prudent to try to solve problems that are not understood.
- It is fruitless to do research whose outputs will not be used in engagement (the social process of informing people).
- Engagement not guided by evidence is usually ineffective and inefficient.

We also found that technical innovation and institutional innovation usually complement each other. At times they are so interrelated as to be codependent. We provided a number of examples.

The experience of the CPWF is that using systems of innovation in a longer time frame is often the key to progress and ultimate success. Innovation is a social process that is risky and unpredictable, where success comes from learning from failure. New ideas are tested, accepted or discarded in a dynamic and iterative process of learning selection. Researchers will be well served if they can foresee the trajectory of innovation. This allows them to build on past learning and to link up with concurrent projects on similar issues being managed by other partners.

Notes

1   See cgspace.cgiar.org/handle/10568/2983 (accessed 14 April 2014).

Acknowledgments

We are grateful for helpful comments and suggestions from Tilahun Amede, Kim Geheb, Elizabeth Humphries, Nancy Johnson and an anonymous reviewer.

References

Alvarez, S., Douthwaite, B. Thiele, G., Mackay, R., Córdoba, D. and Tehelen, K. (2010) 'Participatory impact pathways analysis: a practical method for project planning and evaluation', *Development in Practice*, vol. 20, no. 8, pp. 946–958.

Andreini, M., Schuetz, T., Senzanje, A., Rodriguez, I., Andah, W., Cecchi, P., Boelee, E., van de Giesen, N., Kemp-Benedict, E. and Liebe, J. (2010) *Small multi-purpose reservoir ensemble planning*, CPWF Project Report Series PN46, CGIAR Challenge Program on Water and Food, Colombo, Sri Lanka.

Andreini, M., Schuetz, T. and Harrington, L. (2009) *Small reservoirs toolkit*, CGIAR Challenge Program on Water and Food, www.smallreservoirs.org/ (accessed 14 April 2014).

Asante, S. (2011) *Empowering farming communities in northern Ghana with strategic innovations and productive resources in dryland farming*, CPWF Project Report Series PN06, CGIAR Challenge Program on Water and Food Colombo, Sri Lanka.

Boelee, E., Cecchi, P., and Koné, A. (2009) *Health impacts of small reservoirs in Burkina Faso*, IWMI Working Paper 136, IWMI, Colombo, Sri Lanka.

Bouman, B. A. M. (2008) *Developing a system of temperate and tropical aerobic rice in Asia (STAR)*, CPWF Project Report Series PN16, CGIAR Challenge Program on Water and Food, Colombo, Sri Lanka.

Castillo, E. G., Tuong, T. P., Ismail, A. M. and Inubushi, K. (2007) 'Response to salinity in rice: Comparative effects of osmotic and ionic stresses', *Plant Production Science,* vol. 10, no. 2, pp. 159–170.

Castro, A., Rivera, M., Ferreira, O., Pavón, J., García, E., Amézquita, E., Ayarza, M., Barrios, E., Rondón, M., Pauli, N., Baltodano, M. E., Mendoza, B., Wélchez, L. A. and Rao I. M. (2009) *Quesungual slash and mulch agroforestry system (QSMAS): Improving crop water productivity, food security and resource quality in the subhumid tropics,* CPWF Project Report Series PN15, CGIAR Challenge Program on Water and Food, Colombo, Sri Lanka.

Chowdhury, A. H., Odame, H. H. and Hauser, M. (2010) 'With or without a script? Comparing two styles of participatory video on enhancing local seed innovation system in Bangladesh', *Journal of Agricultural Education and Extension,* vol. 16, no. 4, pp. 355–371.

CIMMYT (1999) 'Risk management for maize farmers in drought-prone areas of Southern Africa', *Proceedings of a Workshop on Maize in Drought-Prone Areas,* Kadoma Ranch, Zimbabwe, 1–3 October 1997, CIMMYT, ICRISAT and DANIDA.

CIMMYT (2010) *Conservation agriculture for the dry-land areas of the Yellow River basin: Increasing the productivity, sustainability, equity and water use efficiency of dry-land agriculture, while protecting downstream water users,* CPWF Project Report Series PN12, CGIAR Challenge Program on Water and Food, Colombo, Sri Lanka.

CPWF (2012a) *Abandoning slash and burn for slash and mulch in Central America's drought-prone hillsides,* Outcome Stories Series, CGIAR Challenge Program on Water and Food, Colombo, Sri Lanka.

CPWF (2012b) *Addressing public health issues in urban vegetable farming in Ghana,* Outcome Stories Series, CGIAR Challenge Program on Water and Food, Colombo, Sri Lanka.

CPWF (2012c) *Aerobic rice: A new crop to help farmers cope with scarce water,* Outcome Stories Series, CGIAR Challenge Program on Water and Food, Colombo, Sri Lanka.

CPWF (2012d) *Improving food security and livelihoods at the interface between fresh and saline water in Bac Lieu, Vietnam,* Outcome Stories Series, CGIAR Challenge Program on Water and Food, Colombo, Sri Lanka.

CPWF (2012e) *Opportunity in adversity: Collective fish culture in the seasonal floodplains of Bangladesh,* Outcome Stories Series, CGIAR Challenge Program on Water and Food, Colombo, Sri Lanka.

CPWF (2012f) *Paying for environmental services in an Andean watershed: Encouraging outcomes from conservation agriculture,* Outcome Stories Series, CGIAR Challenge Program on Water and Food, Colombo, Sri Lanka.

CPWF (2012g) *Vast saline lands reclaimed by simple technologies in coastal and inland Asia,* Outcome Stories Series, CGIAR Challenge Program on Water and Food, Colombo, Sri Lanka.

Cullen, B. (2013) *Communal grazing land management in Fogera: Lessons from innovation platform action research,* Nile Basin Development Challenge–Rainwater Management for Resilient Livelihoods, nilebdc.org/2013/01/18/ip-fogera-lessons/ (accessed 14 April 2014).

Department of Agricultural Extension (DAE), Ministry of Agriculture, Forestry and Fisheries (MAFF), Cambodia and CGIAR Challenge Program on Water and Food (CPWF) (2012) *Commune agroecosystem analysis in Cambodia: A guidance manual.* Ministry of Agriculture, Forestry and Fisheries, Department of Agricultural Extension (DAE), Phnom Penh, Cambodia and CGIAR Challenge Program on Water and Food (CPWF), Colombo, Sri Lanka, 118p.

Dimes, J., Twomlow, S., Rusike, J., Gerard, B., Tabo, R., Freeman, A. and Keatinge, J. D. H. (2005) 'Increasing research impacts through low-cost soil fertility management options for Africa's drought-prone areas, Sustainable Agriculture Systems for the Drylands', *Proceedings of the International Symposium for Sustainable Dryland Agriculture Systems*, 2–5 December, 2003. Niamey, Niger.

Douthwaite, B. (2002) *Enabling innovation: A practical guide to understanding and fostering technological change*, Zed Books, London.

Douthwaite, B. and Gummert, M. (2010) 'Learning selection revisited: How can agricultural researchers make a difference?' *Agricultural Systems*, vol. 103, no. 5, pp. 245–255.

Escobar, G. and Estrada, R. D. (2011) *Diversity of water-based benefits in the High Andes Range*, AN1 Working Paper 1, RIMISP.

Estrada, R. D., Quintero, M., Moreno, A. and Ranvborg, H. M. (2009) *Payment for environmental services as a mechanism for promoting rural development in the upper watersheds of the tropics*, CPWF Project Report Series PN22, CGIAR Challenge Program on Water and Food, Colombo, Sri Lanka.

Fisher, M. and Cook, S. (2012) *Water, food and poverty in river basins: Defining the limits*. Routledge, London.

Gomez-Macpherson, H., Maatougui, M., Kidane, A. and Ghebretatios, I. (2009) 'Water productivity of temperate cereals in Eritrea: Complementing the participatory breeding programme', *Proceedings of the CGIAR Challenge Program on Water and Food International Workshop on Rainfed Cropping Systems,* Tamale, Ghana, 22–25 September 2008, edited by E. Humphreys and R. Bayot, Challenge Program on Water and Food, Colombo, Sri Lanka.

Grando, S., Ghebretatios, I., Amlesom, S., Ceccarelli, S., Maatougui, M. H. el, Niane, A. A., Mustafa, Y., Sarker, A., Abdalla, O., Berhane, T., Abraha, N., Tsegay, S., Isaac, T., Tesfamichael, E., Hailemichael, S., Semere, M., Tafere, T., Araia, W., Haile, A., Mesfin, S. and Ghebresselassie, T. (2010) *Water productivity improvement of cereals and foods legumes in the Atbara Basin of Eritrea*, CPWF Project Report Series PN02, CGIAR Challenge Program on Water and Food, Colombo, Sri Lanka.

Harrington, L. and Hobbs, P. (2009) 'The rice–wheat consortium and the Asian Development Bank: A history', in J. K. Ladha, Y. Singh, O. Erenstein, and B. Hardy (eds), *Integrated crop and resource management in the rice–wheat system*, International Rice Research Institute, Los Baños, Philippines.

Humphreys, E. (2012) *Project Ganges 2: Productive, profitable and resilient agriculture and aquaculture systems*, CPWF Six-Monthly Project Reports for April 2012–September 2012, CGIAR Challenge Program on Water and Food, Colombo, Sri Lanka.

Islam, M. R., Salam, M. A., Bhuiyan, M. A. R., Rahman, M. A. and Gregorio, G. B. (2008) 'Participatory variety selection for salt tolerant rice', *International Journal of Biologial Research*, vol. 4, no. 3, pp. 21–25.

Ismail, A.M. (2009) *Development of technologies to harness the productivity potential of salt-affected areas of the Indo-Gangetic, Mekong, and Nile River basins*, CPWF Project Report Series PN07, CGIAR Challenge Program on Water and Food, Colombo, Sri Lanka.

IWMI (2009) *PN51 Completion Report—Waste water irrigation—opportunities and risks*, CGIAR Challenge Program on Water and Food Colombo, Sri Lanka.

Johnson, N., Garcia, J., Rubiano, J. E., Quintero, M., Estrada, R. D., Mwangi, E., Morena, A., Peralta, A. and Granados, S. (2009) 'Water and poverty in two Colombian watersheds', *Water Alternatives*, vol. 2, no. 1, pp. 34–52.

Kibret, S., McCartney, M., Lautze, J. and Jayasinghe, G. (2009) *Malaria transmission in the vicinity of impounded water: Evidence from the Koka Reservoir, Ethiopia,* Research Report 132, IWMI, Colombo, Sri Lanka.

Kumwenda, J., Waddington, S., Snapp, S., Jones, R. and Blackie, M. (1996) *Soil fertility management research for the maize cropping systems of smallholders in Southern Africa: A review,* NRG Natural Resources Group Paper 96/02, CIMMYT, El Batan, Mexico.

Lebel, L., Bastakoti, R. C. and Daniel, R. (2010) *Enhancing multi-scale Mekong water governance,* CPWF Project Report Series PN50, CGIAR Challenge Program on Water and Food, Colombo, Sri Lanka.

Low, A., Seubert, C. and Waterworth, J. (1991) *Extension of on-farm research findings: Issues from experience in Southern Africa,* CIMMYT Economics Working Paper 91/03, CIMMYT, Mexico City.

Ludi, E., Belay, A., Duncan, A., Snyder, K., Tucker, J., Cullen, B., Belissa, M., Oijira, T., Teferi, A., Nigussie, Z., Deresse, A., Debela, M., Chanie, Y., Lule, D., Samuel, D., Lema, Z., Berhanu, A. and Merrey, D. (2013) *Rhetoric versus realities: A diagnosis of rainwater management development processes in the Blue Nile basin of Ethiopia,* CPWF Research for Development (R4D) Series 05, CGIAR Challenge Program on Water and Food, Colombo, Sri Lanka.

McCartney, M., Gichuki, F. N., Nguyen-Khoa, S. and Kodituwakku, D. C. (2009) 'Living with dams: Managing the environmental impacts', *Water Policy*, vol. 11, Supplement 1, pp. 121–139.

McCartney, M. P. (2009) *Improved planning of large dam operation: Using decision support systems to optimize livelihood benefits, safeguard health and protect the environment,* CPWF Project Report Series PN36, CGIAR Challenge Program on Water and Food Colombo, Sri Lanka.

Merrey, D. and Gebreselassie, S. (2011) *Promoting improved rainwater and land management in the Blue Nile (Abay) Basin of Ethiopia,* NBDC Technical Report 1, ILRI, Nairobi, Kenya.

Mikhail, M. and Yoder, R. (2008) *Multiple use water service implementation in Nepal and India: Experience and lessons for scale-up*: International Development Enterprises, CGIAR Challenge Program on Water and Food, and International Water Management Institute.

Molle, F., Foran, T. and Kakonen, M. (2009) *Contested waterscapes in the Mekong Region: Hydropower, Livelihoods and Governance,* Earthscan, London.

Mukherji, A. (2008) 'Poverty, groundwater, electricity and agrarian politics: Understanding the linkages in West Bengal', *Transforming India* (January 2008).

Mukherji, A. (2012a) 'On winning the Norman Borlaug Award for Field Research and Application', *Waterscapes,* aditimukherji.wordpress.com/2012/09/27/on-winning-the-norman-borlaug-award-for-field-research-and-application/ (accessed 14 April 2014).

Mukherji, A. (2012b) *Project Ganges 3: Water governance and community-based management,* CPWF Six-Monthly Project Reports for April 2012–September 2012, CGIAR Challenge Program on Water and Food, Colombo, Sri Lanka.

Mukherji, A., Shah, T. and Bannarjee, P. S. (2012) 'Kick-starting a second Green Revolution in Bengal', *Economic & Political Weekly*, vol. XLVII, no. 18, pp. 27–30.

Pavon, J., Amezquita, E., Menocal, O., Ayarza, M. and Rao, I. M. (2006) 'Application of QSMAS principles to drought-prone areas of Nicaragua: Characterization of soil chemical and physical properties under traditional and QSMAS validation plots in La Danta watershed in Somotillo', *CIAT Annual Report 2006: Integrated soil fertility management in the tropics*, CIAT-TSBF, Cali, Colombia.

Perrin, B. (2002) 'How to—and how not to—evaluate innovation', *Evaluation*, vol. 8, no. 1, pp. 13–28.

Quintero, M. (2012) *Project Andes 2: Assessing and anticipating the consequences of introducing benefit sharing mechanisms (BSMs)*, CPWF Six-Monthly Project Reports for April 2012–September 2012, CGIAR Challenge Program on Water and Food, Colombo, Sri Lanka.

Quintero, M., Wunder, S. and Estrada, R. D. (2009) 'For services rendered? Modeling hydrology and livelihoods in Andean payments for environmental services schemes', *Forest Ecology and Management*, vol. 258, no. 9, pp. 1871–1880.

Sellamuttu, S. S. (2012) *Project Mekong 1: Optimizing reservoir management for livelihoods*, CPWF Six-Monthly Project Reports for April 2012–September 2012, CGIAR Challenge Program on Water and Food, Colombo, Sri Lanka.

Sellamuttu, S. S., Mith, S., Chu Thai, H., Johnston, R., Baran, E., Dubois, M., Soeun, M., Craig, I., Nam, S. and Smith, L. (2010) *Commune agroecosystem analysis to support decision making for water allocation for fisheries and agriculture in the Tonle Sap wetland system*, CPWF Project Report Series PN71, CGIAR Challenge Program on Water and Food, Colombo, Sri Lanka.

Sharifullah, A., Tuong, T. P., Mondal, M. and Franco, D. (2008) 'Increasing crop water productivity through optimizing the use of scarce irrigation water resources', *Proceedings of the CGIAR Challenge Program on Water and Food 2nd International Forum on Water and Food, Addis Ababa, Ethiopia, 10–14 November 2008*, CGIAR Challenge Program on Water and Food, Colombo, Sri Lanka

Sheriff, N., Joffre, O., Hong, M. C., Barman, B., Haque, A. B. M., Rahman, F., Zhu, J., Nguyen, H. van, Russell, A., Brakel, M. van, Valmonte-Santos, R., Werthmann, C. and Kodio, A. (2010) *Community-based fish culture in seasonal floodplains and irrigation systems*, CPWF Project Report Series PN35, CGIAR Challenge Program on Water and Food, Colombo, Sri Lanka.

Siambi, M. (2011) *Increased food security and income in the Limpopo Basin through integrated crop, water and soil fertility options and public–private partnerships*, CPWF Project Report Series PN01, CGIAR Challenge Program on Water and Food, Colombo, Sri Lanka.

Tripp, R. and Woolley, J. (1989) *The Planning Stage of On-Farm Research: Identifying Factors for Experimentation*, CIMMYT and CIAT, Mexico D.F. and Cali, Colombia.

Tuong, T. P. and Hoanh, C. T. (2009) *Managing water and land resources for sustainable livelihoods at the interface between fresh and saline water environments in Vietnam and Bangladesh*, CPWF Project Report Series PN10, CGIAR Challenge Program on Water and Food, Colombo, Sri Lanka.

Van Hoeve, E. and van Koppen, B. (2005) 'Beyond fetching water for livestock: A gendered sustainable livelihood framework to assess livestock-water productivity', *Nile Basin Water Productivity: Developing a shared vision for livestock production Workshop of the CGIAR Challenge Program on Water and Food (CPWF)*, Kampala, Uganda.

van Koppen, B., Smits, S. and Mikhail, M. (2009a) 'Homestead- and community-scale multiple-use water services: Unlocking new investment opportunities to achieve the Millennium Development Goals', *Irrigation and Drainage*, vol. 58, pp. 73-86.

van Koppen, B., Smits, S., Moriarty, P., de Vries, F. P., Mikhail, M. and Boelee, E. (2009b) *The multiple-use water services (MUS) project,* CPWF Project Report Series PN28, CGIAR Challenge Program on Water and Food, Colombo, Sri Lanka.

van Rooyen, A. and Homann-Kee Tui, S. (2008) 'Enhancing incomes and livelihoods through improved farmers' practices on goat production and marketing', *Proceedings of a workshop organized by the Goat Forum*, 2–3 October 2007, Bulawayo, Zimbabwe.

van Rooyen, A. and Homann-Kee Tui, S. (2009) 'Promoting goat markets and technology development in semi-arid Zimbabwe for food security and income growth', *Tropical and Subtropical Agroecosystems*, vol. 11, pp. 1–5.

van Rooyen, A. (2012) *Project Limpopo 3: Farm systems and risk management,* CPWF Six-Monthly Project Reports for April 2012–September 2012, CGIAR Challenge Program on Water and Food, Colombo, Sri Lanka.

Waddington, S. R., Murwira, H. K., Kumwenda, J. D. T., Hikwa, D. and Tagwira, F. (1998) *Soil fertility research for maize-based farming systems in Malawi and Zimbabwe,* CIMMYT, Mexico, D. F.

Ward, P., Rasaenan, T., Meynell, P. J., Ketelsen, T., Sioudom, K. and Carew-Reid, J. (2012) 'Flood control challenges for large hydroelectric reservoirs: Example from Nam Theun-Nam Kading basin in Lao PDR', *2nd Mekong Forum on Water, Food and Energy,* November 13–15, 2012, Hanoi, Vietnam.

6 Research on institutions for agricultural water management under the CGIAR Challenge Program on Water and Food

Nancy Johnson,[a][*] *Brent M. Swallow*[b] and *Ruth Meinzen-Dick*[a]

[a]International Food Policy Research Institute IFPRI, Washington, DC, USA; [b]University of Alberta, Edmonton, Canada; [*]Corresponding author, n.johnson@cgiar.org.

Introduction

Experience in the past 30 years of water management has shown that technology alone is not sufficient to reduce poverty, enhance food security and increase rural livelihoods. Appropriate institutions are necessary for technologies to be taken up and used, especially in agriculture and natural resource management. In this chapter we review and summarize the findings of the research that was carried out in the Challenge Program on Water and Food (CPWF) on institutions for water management. We also examine how research was used by policy and other decision-makers to influence development outcomes.

We first define water management institutions and present a framework for institutional analysis in a river basin context. Based on a review of peer-reviewed publications, we then summarize what CPWF research projects learned about how water management institutions work and how they can be strengthened. Next we analyze seven projects that translated their research into action on the ground, and identify factors that contributed to their success. Finally, we draw conclusions and discuss the implications.

What are water management institutions?

Institutions may be defined as "the rules of the game in a society or, more formally, the humanly devised constraints that shape human interaction" (North, 1990, p. 1). As such, water management institutions shape the expectations and incentives of various actors, and hence their behavior regarding water use and water-related infrastructure and land management.

Institutions therefore play crucial roles in water allocation, coordination of action, risk management, conflict resolution and overall water governance.

There are many different types of institutions, both formal and informal, that can serve these functions. These can be broadly classified as state, market and collective action (or "customary") institutions (Uphoff, 1993). This classification focuses on the agencies involved in implementation. For example, flood risk may be managed by state intervention, by private-sector markets for flood insurance or by neighborhood assistance—or a combination of these.

Figure 6.1 illustrates the importance of two types of key institutions for agricultural water management. The vertical axis shows the spatial scale of a technology, from an individual plot, through a whole farm, to a community, region and even global scale. All approaches that are above the scale of the individual farm require some form of coordination—either by local organizations, the state or the market. For example, a homestead fish pond or well may be owned by an individual small farmer. But where holdings are small and tubewells have large capacity, farmers may join together to buy and operate a tubewell, or the state may install and operate it, or one farmer can install it

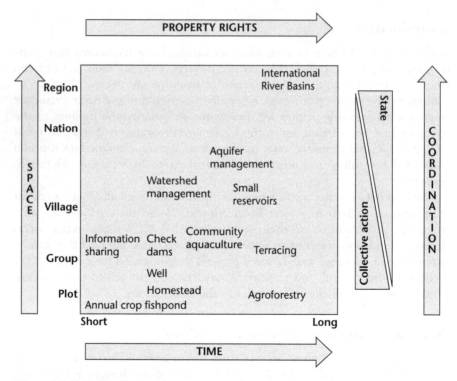

Figure 6.1 Coordination and property rights institutions for agricultural water management.

Source: adapted from Knox et al., 1998.

and sell water to neighbors. Even if a tubewell can be operated independently by one farm household, others might be affected by aquifer depletion; some form of institution is therefore needed to control these externalities.

The horizontal axis indicates the permanence of a technology or approach, or the time frame to cover the investment. The longer the temporal scale, the greater the need for property rights to provide authorization and incentive to make the investment. Even a tenant farmer or a wife without independent rights to her husband's land can install a drip kit, but may not be allowed to install a tubewell, and may not have the incentive to install and maintain terracing or drainage systems for salinity control. Technologies that are not as tied to the land may be more viable for those with insecure land rights. For example, a treadle pump or small motorized pump that can be moved can be used by a tenant. Even if farmers have secure rights to the land, they may not be willing to invest in irrigation systems if they do not also have secure rights to the water (which is often separate from the land). Property rights over water, land and infrastructure may also derive from and be backed by a range of institutions. In many cases, water rights become operationalized through organizations such as water user associations or producer groups. Ensuring that women, smallholders, livestock keepers or other poor and marginalized water users are represented in those organizations is an important step to strengthening their water rights.

Which institution is most appropriate depends on the particular conditions. In general, the advantages of the state are greatest at the largest scale; collective action works at more localized levels. Markets are highly variable in whether they provide effective coordination among smallholders.

As part of CPWF research, Swallow et al., 2006, articulated a conceptual framework for analyzing the performance of institutions in watershed management based on the institutional analysis and development (IAD) framework (Ostrom, 2005; Di Gregorio et al., 2008). The IAD framework begins with the characteristics of the resource, of the user group, and the rules in use. In the watershed context, these include water and financial resources, risk, local and customary institutions, and the high-level institutions that link different parts (upper, middle and lower) of watersheds or provide governance to whole basins (Figure 6.2). These contextual features influence the "action arena" in which various groups of people draw upon their action resources according to the various rules in use. Action resources are tangible and intangible assets and personal characteristics that enable people to take action or influence others' decisions. The interplay of the people, rules and action resources results in individual and collective action that shapes patterns of interaction, especially land use and water resource investments. These patterns lead to outcomes and effects on welfare and water resource conditions, which in turn feed back to the context and action arenas in the future.

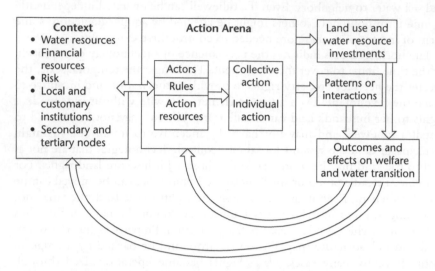

Figure 6.2 The framework for institutional analysis applied to multi-scale watershed management.

Source: CPWF, 2013.

What did CPWF learn about water management institutions?

Publications

To keep the task of synthesis manageable, we focused our analysis on peer-reviewed publications. We identified and accessed a total of 68 such publications (see chapter appendix, Table 1), including 6 books, 6 book chapters, 54 journal articles and 2 working papers suitable for review.[1] We were as systematic as possible and believe the publications are representative of the research done in CPWF in terms of methods, approaches and findings.

The research was about institutions for water management, so it is interesting to note the range of thematic areas and disciplines represented by the journals (chapter appendix, Table 2). Environmental sciences was the commonest subject (19), followed by social science (8), agricultural and biological sciences (6), business, management and accounting (4), earth and planetary studies (4), economics (2), and engineering (2). Nearly all the journals are multi-disciplinary.

The unweighted mean of the impact factors for the 27 journals that published impact factors was 1.53. A Google Scholar analysis of the citations of those articles[2] and book chapters on April 18, 2013, showed that the papers were cited a total of 1561 times, for an average of 23 citations per paper. A review paper that compared governance in the Mekong to other complex governance situations (Lebel et al., 2006) was cited 346 times. Seven papers were cited more than 50 times, nine were cited 20–50 times, fifteen 11–20 times, and the remainder were cited 1–10 times. These citation numbers suggest a large impact of the CPWF on the scientific community concerned with institutions, especially given since the average time since publication was only 5.2 years.

The list of authors of CPWF papers on institutions shows the convening power of the CPWF. Many of the papers were authored by scientists at CGIAR Centers other than International Water Management Institute (IWMI) (e.g. Centro Internacional de Agricultura Tropical (CIAT), International Food and Policy Research Institute (IFPRI), World Agroforestry Centre (ICRAF) and by leading scholars in their respective fields. Google Scholar shows that Louis Lebel is widely cited in the field of resilience science and governance of complex systems and Juan Camilo Cardenas is a leading scholar in the field of experimental economics.

Synthesis through the lens of the IAD framework

We used the IAD framework (Figure 6.2) to synthesis the information on publications. They fell into three main groups. The first group focused on *better understanding of action arenas*. They asked questions such as, "What institutions influence water management in a given context?" or, "What factors determine people's ability to have influence in a given action arena?" The second group explored *how institutions affect water allocation*, and what implications this has for welfare and environmental outcomes. Questions asked were: "How do alternative policies and institutions affect individual water-use decisions, collective water-use outcomes, or the returns to alternative water uses?" and "What are the welfare and environmental implications of alternative allocations?" The third group looked at *institutional innovation processes*, asking questions such as, "How do institutions evolve and how can research-for-development (R4D) interventions strengthen the participation of poor or marginalized groups in water management institutions and processes?"

Some papers fitted in more than one category, but we classified them where they seemed to fit best. A few of the papers did not fit in any of the categories and were removed from the analysis.

Action arenas

The concept of action arenas can be applied at different levels, for the CPWF portfolio, ranging from the household to the transnational river basin. Lebel et al. (2009) examined the role of aquaculture in affecting bargaining power between men and women at the household and community levels. They found that participation in aquaculture provided action resources that empowered women within the household, but only contributed indirectly to women's empowerment at the community level. Leadership in women's groups was more likely to provide action resources that translated into authority in the community action arenas.

Watersheds were a more common action arena for water-related institutional studies. Johnson et al. (2009), Lebel and Daniel (2009) and Ma et al. (2008) discuss interactions among communities, governments and firms in negotiating the trade-offs between environmental conservation and economic development, as well as the sharing of risks and benefits. All papers acknowledge that people and the rules that govern decision-making affected outcomes. Lebel and Daniel (2009) note that multi-stakeholder planning processes can lead to better outcomes; but they do not eliminate power relations, which points to the need to understand the action resources that different parties hold, and how these may be deployed to affect watershed decision-making. The study by Johnson et al. (2009) of two watersheds in Colombia found that the trade-offs between environmental conservation and poverty reduction are site-specific. The interests of rich and poor were not always in conflict, especially when there was diversification of livelihoods to include off-farm options.

The environment-livelihoods' trade-offs play out in different ways when fisheries are considered, again depending on the context. In coastal zones, Gowing et al. (2006) examined the (unplanned) action arena where agriculture, shrimp farming and fishing conflicted. Shrimp farmers had taken advantage of ambiguous resource tenure to expand their farms at the expense of other livelihood activities. In the Mekong, Friend and Blake (2009) examined the trade-offs between capture fisheries and hydropower development, and how these played out in policy choices. Hydropower advocates had stronger political influence because the development of their sub-sector fitted with dominant development pathways and paradigms. Evidence on the value of production from capture fisheries provided a potential action resource for fisheries advocates, but Friend and Blake (2009) argued that the evidence will not be effective unless there is broadening of the decision-making arena.

Many studies pointed to the importance of action arenas that allow multi-stakeholder participation in decision-making (Molle et al., 2009). But Sajor and Thu's (2009) analysis of the Saigon River showed that top-down processes can negate participation, even if there are formal provisions for it.

Action arenas are not just about decision-making—there is also a need for enforcement. Hagos et al. (2011) point to a lack of institutional enforcement capacity for land use and pollution control in the Blue Nile sub-basins. Regulations from formal institutions with command and control approaches argued for building on existing institutions with local or self-enforcement. Manuta et al. (2006) examined the institutional incapacities that limit the effectiveness of flood disaster management in Southeast Asia. Hagos et al. (2011), Manuta et al. (2006) and Sajor and Ongsakul (2007) showed that administrative fragmentation shifted environmental costs on to the poor. Integrated water resources management (IWRM) calls for multi-stakeholder processes to deal with the multifaceted nature of water management. But action arenas in which the poor were able to participate effectively, and where decisions could be enforced, did not emerge spontaneously. Analyzing the development of the Red River Basin Organization in Vietnam as a vehicle for IWRM, Molle and Hoanh (2009, p. 14) note, "Management regimes require bureaucratic configurations, legal frameworks and governance patterns that are consistent with these regimes. Pushing for a particular regime when these conditions are not met may just be wishful thinking with little chance of success."

International transboundary water governance was the highest-level action arena examined by the CPWF. Lautze and Giordano (2005, 2007a and b) reviewed transboundary water laws and agreements in Africa. They found that they tended to follow western models in prioritizing environmental protection over economic development of water resources. This was despite the low overall level of water resource development in Africa, and the pressing need for increased food security and economic development. They attributed this bias to imported models of transboundary water law to developing country contexts rather than developing institutional arrangements designed to fit specific local situations.

Merrey (2009) argued that the various commissions and authorities that were set up to promote integrated water resource management were not operationalized and so were ineffective. This is because they were models, rather than being based on indigenous institutional principles that were more suited to African conditions.

Manzungu et al. (2009) linked the ineffectiveness of Botswana's policies for water management for livestock to imported models of modernization. This was instead of having the state work with local communities to develop a policy that is adapted to conditions in Botswana.

Lebel et al. (2005) pointed out that the scale of action arenas at which policies and water management decisions are made affects the power and action resources available to different people.

> Scale choices can be a means of inclusion or exclusion . . . People, institutions, and landscapes are made to fit levels and scales in the states' systems of accounting and monitoring. Local-level knowledge and institutions are seen as local in scope, relevance, and power, whereas the rules and knowledge of the state have much bigger scope and significance.
>
> (Lebel et al., 2005)

Building on local knowledge and institutions, instead of importing external ones was likely to lead to patterns of interaction that were more suited to local conditions, but putting this principle into practice remained a challenge.

Role of institutions in water allocation

Most of the papers in this category focused on policies. They either assessed the impacts of specific policies on water allocation or developed methods to assess impacts and trade-offs associated with alternative policy scenarios.

A small number looked at factors that determined the demand for and performance of community-based water systems. Others looked at the potential for changing water allocation through schemes for payment for environmental services.

Policy's positive role was highlighted in several papers. Xu et al. (2007) documented the transition from forest in Yunnan Province in southwest China over a 50-year period. They provided empirical evidence that transitions were driven by economic growth that created off-farm opportunities, complemented by state policies favoring conservation. They showed how the dynamic transitions contributed to biodiversity conservation, carbon sequestration and regional economic growth, and provided lessons for the region.

Lautze and Giordano (2006) analyzed a database of transboundary water agreements in Africa. They showed that agreements that specifically mention equity—a growing trend in recent years in response to international pressure— divide water more equitably than agreements that do not mention equity.

Shah et al. (2008) looked at the impact of separating the agricultural and non-agricultural electricity grids on groundwater use in Gujarat in India. Subsidized electricity for pumping was a major cause of groundwater overdraft. It overloaded the electricity grid, resulting in economic and environmental losses. It was politically difficult to reform water allocation directly, but rationing electricity rather than water changed the action arena. It addressed the issue of water for agriculture and generated high net benefits. While some farmers were hurt by the scheme, overall quality of life improved, especially in the non-farm sector.

Tsegaye and Berger (2007) assessed the impact of removing subsidies on irrigated farming in Ghana. The analysis showed that there was a strong complementarity between irrigation farming and off-farm employment, both of which depend on household labor endowment. The complementarity suggests that where credit markets are weak, irrigation farmers generate financial liquidity from off-farm activities, which could lead to larger families in the long run.

When policy changes, there are always winners and losers. Several studies developed methods to measure the value of trade-offs of a variety of policy and decision options, across scales, uses and users, using bioeconomic models.

Baran et al. (2006, 2010) assessed trade-offs between rice farming, aquaculture and the environment in Vietnam using a Bayesian network model. They showed that a mixed strategy of rice and shrimp was better than either rice or shrimp alone. This was consistent with studies that looked at how and why past changes in water management had affected livelihoods and the environment (Gowing et al., 2006; Khiem and Hossain, 2010).

Berger and Schreinemachers (2006) and Schreinemachers and Berger (2006) developed a mathematical programming-based multi agent system (MP-MAS) computer model. It captured interdependencies between individuals and their incentives for cooperation around water management. Berger et al. (2007) developed a decision-support tool to assess the impacts of alternative policy scenarios in river basins in Chile. Bharati et al. (2008) described the development, calibration and application of a dynamic economic-hydrologic simulation model. It was used to evaluate the conjunctive use of surface and groundwater in irrigation systems based on small reservoirs in the Volta Basin in West Africa.

The multi-level scenario analysis approach (Lebel, 2006) is an alternative to detailed bioeconomic models. It was used to explore uncertainties about how livelihoods and landscapes in upper tributary watersheds of Southeast Asia might unfold in the coming decades. It developed and analyzed nested scenarios at local and regional scales. The regional scenarios looked at market and political integration issues while the local scale scenario focused on the resource base and the role local stakeholders could play in their management. The scenarios were intended as a starting point for discussions among stakeholders and as a framework for designing and interpreting simulation studies of land use and change in land cover. They were also used as a tool to identify strategies for resilient livelihoods and regional development.

Several papers looked at the performance of community-based water systems, especially the factors that determine how water is provided and allocated in such systems. The beyond-domestic paradigm highlighted the benefits of integrated approaches to managing water both for agriculture and domestic use in developing countries. Multiple-use water services delivered water for both domestic and productive use in nine countries (van Koppen et al., 2009). Domestic and productive use of homestead systems were economically important, averaging between US$40 and US$80/person/yr. While average benefits were high, they were not always distributed equitably across different types of users.

In South Africa poorer users were only willing to pay for domestic water, while better-served households were also willing to pay for water for productive uses (Kanyoka et al., 2008). In the upper reaches of the Nyando Basin in Kenya, small amounts of water used for vegetables and dairy cattle boosted income by 30 percent (Crow et al., 2012). The benefits varied by type of system and were larger for women than men in some systems. There were pre-conditions that guaranteed successful collective action for community-based systems.

A number of papers explored the potential of payment for ecosystem services (PES) schemes to change the way land and water resources were used. Rubiano et al. (2006) laid out a framework for analyzing alternative land use scenarios. Quintero et al. (2009) applied a bioeconomic model based on the framework to look at the potential of PES from changing agricultural land management in several Andean watersheds. PES was economically, socially and environmentally feasible in some contexts, though the results were not necessarily pro-poor.

Cardenas et al. (2010) confirmed that vertical asymmetries in appropriation—the "head ender–tail ender" problem—in a watershed context reduced cooperation in Colombia and Kenya. They led to less total water available and lower overall social welfare. Institutional innovations such as communication or regulation can improve cooperation in economic experiments, however the results were context specific. Jack (2009) found that in the Kenya watersheds, underlying social norms and preferences had a strong influence on people's decisions. Enforcement mechanisms caused crowding out (replacement of internal motivation by external rules) of social preference for equity, which Cardenas et al. (2010) also found with fines for non-compliance with regulations. High fines reduced cooperation if they generated resentment and if people knew that there was a low chance of being caught violating the regulations.

Beyond the many site-specific findings, the CPWF research on how institutions influence the way that water was allocated had several general lessons. Well-designed policies for improving water management had net positive impacts on both the environment and economic growth. Good understanding of the impact of policy or regulatory change on different users could be the basis for targeting interventions to support vulnerable groups who might be negatively affected. The understanding could also provide incentives for people to participate in broad-based collective action. Several innovative approaches were developed to do this in CPWF. Finally, the findings on PES suggest that while such schemes may be economically and ecologically feasible, careful attention to the social context will be necessary for them to work in practice.

Institutional innovation processes

The final group of papers and the book by Molle et al. (2009) looked at how water management institutions evolve. They also looked at how institutions can be strengthened to lead to more equitable and sustainable outcomes. Lebel et al. (2006) assessed how governance in socio-ecological systems increased resilience. They used case studies to answer the question: How do certain attributes of governance function to enhance resilience? Three specific propositions were explored: (1) participation builds trust, and deliberation leads to the shared understanding needed to mobilize and self-organize; (2) poly-centric and multilayered institutions improve the fit between knowledge, action and social-ecological contexts in ways that allow societies to adapt better at appropriate levels; and (3) accountable authorities that also pursue just distributions of benefits and involuntary risks enhance the adaptive capacity of vulnerable groups and society as a whole. Lebel et al. (2006) found some support for parts of all three propositions. They concluded that analysts, facilitators, change agents, or stakeholders, need not only ask, "The resilience of what, to what?" but must also ask, "For whom?"

Though Lebel et al. (2006) found support for all three propositions, their first proposition on the importance of participation and deliberation is the easiest for external interventions to address. Many CPWF projects developed or adapted tools and approaches to identify and better analyze stakeholders and to facilitate more systematic, informed and equitable interaction among stakeholders.

NetMap[3] is a tool that combines social network analysis and influence mapping to help understand, visualize, analyze, discuss and improve situations in which many different people influence outcomes (Schiffer and Peakes, 2009). The tool was developed and used in the White Volta Basin in Ghana (Schiffer and Hauck, 2010). The results helped individuals and groups devise strategies and plan their networking activities more effectively. The tool was then used for a wide range of topics where understanding influence and power are important.

Many tools existed for facilitating stakeholder interaction in various development contexts. Several papers documented the experience of adapting and applying those tools to new geographical areas or sectors. Magombeyi et al. (2008) described and analyzed the implementation of a river basin game. They used it as a tool to facilitate negotiations among upstream and downstream irrigation water users in Ga-Sekororo, in the Olifants River basin in South Africa. By improving people's understanding of the catchment itself and the situations of other stakeholders, the group was able to reach more equitable agreements on water use and sharing. Penning de Vries et al. (2007) identified the 11 "cornerstones" that must be in place for the effective implementation of multiple-use water services (MUS). They proposed to use the learning alliance (Lundy, 2004), a mechanism for joint learning in the context of agro-enterprise development, to assess whether the cornerstones are in place in the MUS project sites.

A series of papers describes the adaptation and implementation of the companion modeling (ComMod) approach (Bousquet et al., 2007). ComMod combined participatory role playing with computer simulation modeling to support improved multi-stakeholder decision-making in rural communities. ComMod was used to support water sharing in Bhutan (Gurung et al., 2006), soil and water management in Northern Thailand (Barnaud et al., 2006; Barnaud et al., 2007) and ethnic conflict around the establishment of a national park in Thailand (Ruankaew et al., 2010). ComMod improved information sharing and collective learning about the importance of collaboration for adaptive and sustainable management of resources.

The tools above allow stakeholders to interact in new ways and with new information. Several authors pointed out that stakeholders must understand and be comfortable with the process, the models and the scenario analysis approach (Becua et al., 2008). Managing the interactions between more and less powerful stakeholders was important to achieving good results. Researchers and local champions had important roles to play to help make this happen (Barnaud et al., 2010). In their assessment of the participatory approaches of a small reservoir project in Ghana, Poolman and Giesen (2006) emphasize the need for a thorough stakeholder analysis. It is also important to use adequate and appropriate forms of stakeholder engagement. This is to ensure real participation in both the short and the long term if project teams expect the interventions to continue once the project and its support ends.

Future work should focus on measuring whether implementation of these promising approaches is associated with real changes in decision making and resource use beyond the game or the project context. In most cases, a supportive institutional environment will be important to sustain and scale-up impacts from the implementation of these approaches. This is consistent with Lebel et al.'s (2006) propositions two and three. Many of these approaches are time and resource intensive to implement even at a small scale. It is therefore important to think about how these pilot experiences can be used strategically to support change at other scales and on other aspects of governance. Another issue is what to do where the institutional context is not initially conducive.

How have the results of CPWF research been used?

While all the CPWF's research was expected to generate knowledge relevant to pro-poor development, some of it had the explicit goal of contributing to change on the ground. These projects are examples of boundary work (Clark et al., 2011), referring to the distinct, yet porous, boundaries between science and policy (Guston, 1999). Boundary organizations and boundary agents have a distinct role to span those boundaries and translate between the languages of science and action (Buizer et al., 2010). Boundary objects help to distill scientific results in ways that are meaningful for action (Fujimura, 1992). Overall, research communities undertake boundary work to "organize their relations with new science, other sources of knowledge, and the worlds of action and policy making" (Clark et al., 2011, p 1).

Clark et al. (2011) formalized these observations in the boundary framework. The framework defines uses of knowledge by knowledge consumers, including: (i) better understanding (called "enlightenment" by Clark et al.), which is advancement of general understanding that is not targeted to specific users or specific actions; ii) decision support, which is supporting specific choices by a single user such as a farmer or a government minister; and (iii) negotiation support, which is supporting negotiation, bargaining or political processes that involve multiple users. The framework also posits two distinct sources of knowledge, either a single source or multiple communities of expertise. Boundary work thus entails effective communication and translation of knowledge between sources and users. The simplest boundary work involves a single community of knowledge sharing knowledge to support better understanding. The most complex boundary work involves multiple communities of knowledge sharing knowledge in support of negotiation.

There is a growing body of research that shows that the effectiveness of boundary work depends upon the credibility, salience and legitimacy of the knowledge that is being shared (White et al., 2010). Knowledge that is credible is technically adequate in the way that it handles evidence. An important indicator of credibility is publication of findings in a peer-reviewed publication. Knowledge that is salient is relevant to a decision or policy under consideration; and knowledge that is legitimate is fair, unbiased and respectful of all stakeholders.

Knowledge for better understanding must be judged to be credible, knowledge for decision support must be judged to be both credible and salient, while knowledge used to support negotiations must be considered to be credible, salient and legitimate. Credibility, salience and legitimacy are difficult to achieve for a single community of expertise, and even more difficult to achieve for multiple sources of expertise. Clark et al. (2011) showed that successful boundary spanning in all six areas of knowledge-to-action depends upon effective participation by relevant scientists and decision makers. Mechanisms of accountability must ensure that both scientists and decision makers have meaningful input into the research process. Boundary objects are created jointly to communicate knowledge in ways that can be accurately understood by relevant scientists and decision makers.

The goal of influencing action on the ground may have implications not only for what research is done but also for how it is done and what other activities beyond research a project team undertakes. Therefore, the unit of analysis in this section is research projects rather than publications. This enables us to look at what projects did to span boundaries between sources and uses of knowledge. We do this to increase the likelihood that the results of the research on institutions for water management will be used.

Projects that seek to influence decisions

Projects that seek to influence decisions must be both credible and salient. The projects analyzed are:

- Managing water and land resources for sustainable livelihoods at the interface between fresh and saline water environments in Vietnam and Bangladesh (PN10).
- Environmental services in rural development (PN22).
- Models for implementing multiple-use water services for enhanced land and water productivity, rural livelihoods and gender equity (PN28).

For each project we looked at how the research results, the actions taken by the project members and the characteristics of the projects and project teams influenced the outcomes achieved by the projects. Since few projects were subjected to external evaluations, the evidence of outcomes is, for the most part, reported by the project teams.

Decision-support activities of PN10: Managing water and land resources for sustainable livelihoods at the interface between fresh and saline water environments in Vietnam and Bangladesh

This project conducted both institutional research and technology development and testing. It sought to develop knowledge and tools to help decision makers identify optimal water-management strategies to balance trade-offs between rice farming, shrimp production and the environment in the Mekong delta. Technology work focused on improving productivity of rice in these systems.

In Vietnam, the decision-support system (DSS) developed by the project (Baran et al., 2006; Baran et al., 2010) was targeted at decision makers at the provincial level who made decisions about sluice-gate management to control the level of salinity in the water. Letting in more seawater is good for shrimp production but lowers production for rice farmers. The project reports that officials used the results of the model in their management decisions to the benefit of both groups of users (PN10 final report).

Several factors may account for this successful influence. In terms of the results themselves, the model recommendations did not support the "rice bowl" (emphasis on rice production) orientation of the existing policy. They were not entirely unexpected, however, and served to confirm a growing sense among stakeholders that the current policy was no longer optimal or sustainable.

During implementation, the PN10 project team engaged farmers, practitioners and policymakers. The model was built based on consultations with farmers, primarily to obtain technical parameters but also to confirm how farmer objectives were modeled. The model was originally intended to maximize biomass but shifted to a multi-objective problem on the basis of farmers' feedback. The targeted decision makers were also engaged from the beginning, and the appropriateness of the model was confirmed with them during development of the model.

Understanding the problem and engagement with local stakeholders may have been facilitated by the special composition of the project team. Among the project principal investigators (PIs) were: natural and social scientists, researchers and development practitioners, nationals from each of the countries where the project worked, and scientists from international rice, fish and water management institutes. This was possible in a small team because in many cases a single individual fitted multiple categories. This likely contributed to the extent to which the researchers were able to work effectively across disciplines, scales of analysis and objectives (research and development).

The work in PN10 built on past projects that the PIs had been involved in, and was linked to projects that other partners led. Even though PN10 was led by CGIAR Centers and governed by rules of CPWF contracts, the partnerships were long-standing. There was a sense of mutual accountability for actions, findings and outcomes that went beyond what was stipulated in CPWF contractual arrangements. Also, the particular characteristics of the PIs and their multiple allegiances could have helped ensure equitable interaction among different stakeholders.

While the project appears to have been successful in getting its results used by provincial water management authorities, several questions remain about the sustainability of the impacts. First, it is not clear to what extent the provincial water managers built their capacity to use the model rather than just the results. The real impact of the investment in the DSS will come when the intended users can adapt and use the model routinely. Second, by targeting these decision makers, did the project reinforce a top-down system? According to project accounts, farmers said that they wanted community rather than provincial management of water. The project also reported frankly that though they were invited to participate in focus group discussions, women's voices were not heard, which means that if their opinions differed from those of the men, they were unlikely to have been taken into consideration.

Decision-support activities under PN22: Environmental services in rural development

This project looked at the potential of PES to stimulate economic and social development in the Andes. The project developed a framework (Rubiano et al., 2006) and a bioeconomic model (ECOSAUT) and applied them at sites in the Andes (Quintero et al., 2009). The objective was to quantify the economic, social, and downstream environmental impacts of adopting improved farming and land management in the upper parts of catchments. The project worked with local stakeholders to design, fund and implement PES schemes that were informed by the results of the modeling work but adapted to suit local social and political contexts. Intended users of the results were organizations that would pay for services, for their own use or on behalf of others. The project targeted sites that were ecologically appropriate and where there was a willing payer for environmental services.

The project team was well integrated with local organizations, both public and private, in both the agriculture and environment sectors. In Colombia, the project worked with the environmental authority (CAR) in the Fuquene watershed. As a result of project efforts, a fund was created to support the adoption of conservation agriculture in ecologically sensitive parts of the watershed. In Peru, the project worked with the municipal water agency to identify alternatives to slash-and-burn agriculture. Implementation of the scheme required project partners to get approval from the national government to adjust water charges to enable payments to farmers.

Members of the project team were proactive in terms of presenting innovative options to local decision makers. They did not target their work to the decisions that decision-makers identified as being highest priority, but rather showed decision makers new ways to deal with current problems. This may have worked because the project team members were well-informed about the local situations. They were also highly regarded nationally and internationally—evidenced in part by the support from CPWF—for their expertise in their fields of work.

As with the previous projects, this CPWF project built on a history of other projects involving some of the same individuals and organizations. The project report noted that GTZ, a partner in the project, played a key role as an interlocutor between project and governments (Estrada et al., 2009).

The project showed that PES schemes are feasible but not necessarily pro-poor or linked to local stakeholder concerns and values (Estrada et al., 2009). Work in the Andes under Phase 2 of the CPWF addressed the same issues in a more integrated way that went beyond market-mediated service provision. It looked at governance and equity issues, as well alternative mechanisms to strengthen social organization and participation in the watersheds (Candelo et al., 2008; Johnson et al., 2009; Cardenas et al., 2010).

Decision-support activities under PN28: Models for implementing multiple-use water services for enhanced land and water productivity, rural livelihoods and gender equity

Most rural water systems in developing countries are built for a specific use. Domestic systems are designed to deliver small amounts of higher-quality water for household use. Productive systems such as irrigation systems are designed to deliver large amounts of lower-quality water. This project conducted research (van Koppen et al., 2009) and advocacy around the idea of designing and implementing water systems for multiple (domestic and productive) uses.

This was one of the few projects subjected to an ex post assessment (Merrey and Sibanda, 2008) and the conclusions of this section draw on those findings. According to the assessment, the MUS project was not a typical research project but rather an "advocacy project of a new concept seeking to maximize its impact through joint learning with stakeholders." The project's primary impacts were on global understanding and appreciation of MUS as a concept. But the project also achieved concrete impacts in some of the countries where it worked, including Colombia, Thailand and Nepal.

The best example is in South Africa. Merrey and Sibanda (2008, p. 20) state that while no direct impacts had yet been felt on the ground in terms of changes in water systems or water availability,

> [T]here is evidence that such changes will begin happening in the near future. First, South Africa's Department of Water Affairs and Forestry (DWAF) has produced a draft *Guidelines for Municipalities* (DWAF, 2006) . . . and is exploring how to adapt the *Municipal Infrastructure Grants* . . . to enable implementation of MUS. Second, MUS principles are included in the current draft of the *Strategic Framework for Water for Sustainable Growth and Development* (W4GD; see DWAF [2006]). This is in line with the strong poverty and gender focus of South Africa's water supply policies and renewed focus on user consultation.

Merrey and Sibanda (2008, p. 20) conclude:

> There is therefore no doubt about the impact of the MUS Project on knowledge and awareness in South Africa, confirming the pathway through which the project contributed to impact.

Merrey and Sibanda (2008) attribute the project's success to the experience and reputations of the project partners. The International Water and Sanitation Centre had long-standing engagements in South Africa and contacts with DWAF and the Water Research Commission. The non-governmental organization, Association for Water and Rural Development, which did much of the practical on-the-ground research and advocacy for multiple-use water services, had been a pioneer in MUS before the CPWF Project began. Perhaps the most important factor, however, was "the personal relationships and reputation of the MUS Project Leader (IWMI's Barbara van Koppen); that is, if she were not located in South Africa or did not have these assets, it is unlikely that multiple uses would have progressed to the current level in policy discussions" (Merrey and Sibanda, 2008, p. 20). The support of CPWF and exposure of MUS concepts at the World Water Forum in Mexico added legitimacy, and the close personal and professional relationships with senior DWAF officials—which led to involvement of DWAF in the project from the beginning, was critical to the achievement of the outcomes.

While the South Africa case was unique, the quality and reputation of the project in countries such as Thailand and Colombia was also due in part to the influence of local partners and partner organizations. In terms of process, "[It] is notable also that the project was able to integrate and add value to local approaches to learning, such as the Farmer Wisdom Networks in Thailand and the South African ... approach [of securing water to enhance local livelihoods]" (Merrey and Sibanda, 2008, p. viii). One way that this was done was through MUS Project impact theory, which proved to be a salient guide to using action research as a tool for achieving impacts. The learning alliance concept (Penning de Vries, 2007) was an effective way for different stakeholders to interact and engage in social learning.

Merrey and Sibanda (2008) question whether the project's decisions to focus on advocacy rather than on in-depth action research (because of budget cuts) could undermine impact in the longer run. Advocacy made the case for "why" water policy should incorporate MUS principles, but without better knowledge on "how," it will be difficult to adopt new standards and practices in a timely manner. In Colombia, for example, less impact was achieved on policy at the national level. In addition to case studies, however, there was curriculum development and a diploma course to inform the future professionals who would be charged with implementing MUS.

Despite the diversity of contexts and approaches, several lessons emerge from looking at the factors that enabled these projects to achieve outcomes. We synthesize these around Clark et al.'s (2011) proposed criteria for boundary work, namely that it achieves participation and accountability, and that it produced boundary objects.

Participation in decision support

All three projects succeeded in spanning the boundaries between different sources of knowledge and between sources (researchers) and users (decision makers). In many cases, these links already existed since researchers had worked together before and had previous experiences living in the region (as nationals or expatriates) and working on the issues. Personal connections with policymakers at different levels were important, enhanced in some cases by recognition associated with international organizations (IWMI or CPWF). In some cases the projects were strengthened by these relationships—e.g., researchers and local authorities in Colombia and Peru in PN20.

Facilitating factors were strong mutual understanding, mutual interest and flexibility in the use of resources for activities of mutual benefit. As Clark et al. (2011) conclude, one-off interactions between researchers and policymakers could lead to useful outcomes but the potential for longer-term impact at scale seems less likely. Sustained collegial relations—formal or informal—between senior researchers and policymakers are important for achieving policy influence.

Accountability in decision support

The CPWF hypothesized that mutual accountability in setting the research agenda was critical for policy influence. While research organizations were the project leaders in all cases, CPWF did require letters of support from local partners, which gave them some leverage over the agenda. Planning workshops and regular feedback, formal and informal, were part of the processes for these three projects. More important, however, seems to be the fact that project teams and partners had worked together in the past and would continue to do so in the future. This provided researchers with a strong incentive to make sure that key stakeholders, including policymakers, were on board and were happy with the way work was progressing. The best example of this could be the MUS South Africa Project where it was concluded that despite the fact that IWMI led it, there really was no leader: all partners including DWAF worked together and complemented each other.

Boundary objects in decision support

Two of the three projects produced bioeconomic models that incorporated input from local stakeholders and were intended to support decisions made by local policymakers. Building the model was an opportunity for close interaction, and afterwards researchers provided policymakers with the results (presentations and policy briefs), tools, and training in how to use them. The way the models were used varied across the projects. The results of the model were applied directly in PN10, and were adapted when designing the PES schemes in PN22.

In the MUS project (PN28), the boundary objects were case studies, the impact pathway approach and the learning alliance. The extent to which these objects contributed to the policy influence, however, is not clear. They took place in all project countries while outcomes occurred in only a few, with a major influence on policy only in South Africa.

Projects that seek to support negotiation

Research that seeks to support negotiation must be seen as credible, salient and legitimate by potential users. We identified three Phase 1 CPWF projects that appeared to embrace the negotiation–support approach:

- Enhancing multi-scale Mekong water governance (PN50);
- Companion modeling and water dynamics (PN25); and
- Sustaining collective action that links across economic and ecological scales (SCALES) in Colombia (PN20).

Negotiation-support activities of PN50

The Mekong Program on Water, Environment and Resilience (M-POWER) was established in 2004. It was a knowledge network undertaking action-based research, facilitated dialogues and knowledge networking to improve water governance in the Mekong region. At least two CPWF projects, PN50, Enhancing Multi-Scale Mekong Water Governance and PN67 Improving Water Allocation were established as keystone projects of M-POWER. The companion modeling project (PN25 described below) was also implemented in the Mekong region.

The goal of PN50 was to improve livelihood security, human health and ecosystem health in the Mekong region through democratizing water governance. Through PN50, the M-POWER consortium implemented an ambitious program of research and direct engagement with stakeholders in the Mekong region. M-POWER defined the Mekong region by the drainage areas of the Southeast Asian rivers: Irrawaddy (most of Burma/Myanmar), Salween (parts of China, Myanmar/Burma, Thailand), Chao Phraya (most of Thailand), Mekong (parts of China, Myanmar/Burma, Laos, Thailand, Cambodia and Vietnam) and Red (parts of China and Vietnam).

For M-POWER, water governance refers to the "ways in which society shares power with respect to decisions about how water resources are to be developed and used." Democratization of water governance "encompasses public participation and deliberation, separation of powers, accountability of public institutions, social and gender justice, protection of rights, representation, decentralizations, and the dissemination of information" (Lebel et al., 2010, p. 17). Little of the research conducted under PN50 focused on water as such; instead most research focused on the structure and function of institutions and power relations. The project engaged with a wide range of actors involved with water resource management in the Mekong region. It provided information to support negotiations and social learning related to fisheries, flood control, hydropower, land and water management in the region. M-POWER engaged in various multi-stakeholder dialogues at the regional level and at the local level in several distinct locations. One of the most important M-POWER contributions was to negotiate support to convene the regional meeting Mekong region waters dialogue: Exploring water futures together.

The final report for PN50 (Lebel et al., 2010) claims the following positive impacts regarding negotiation support: (1) strengthening local representation into planning and implementation; (2) improving the quality of deliberative processes; and (3) enhancing the constructive interplay between state and non-state actors at various levels. The report also acknowledges a central constraint in the Mekong region that has limited progress toward the project goals. Dominant political structures in the region vary from authoritarian, single-party to semi-democracies. In this context, democratization of water governance is seen as a threat to those established powers (Lebel et al., 2010, p. 15).

Negotiation support activities of PN25

The title of PN25 is Companion modeling for resilient water management: Stakeholders' perceptions of water dynamics and collective learning at the catchment scale (ComMod). The objectives of the project were: (1) develop multi-agent simulation tools for facilitating collective assessment of water management problems; (2) build capacity to apply those tools; and (3) participatory construction of concrete propositions to increase water productivity. ComMod was implemented in three upper and three lower sites of the Mekong Basin and in three upper catchments in Himalayan highlands in Bhutan. A range of water management challenges were encountered in those nine project sites. The ComMod project specifically focused on the challenge of integrating multiple sources of knowledge into research for action in those different contexts, addressing the following questions: (1) how to model different stakeholders' perceptions; (2) how to integrate indigenous and science-based knowledge to create a common representation of the system; and (3) how to use models of multi-stakeholder decision-making to improve water management at the catchment scale (Governance Author Team, 2010, pp. 3–10).

> ComMod used conceptual models, role-playing games and agent-based models and simulations in an iterative way, alternating field and laboratory activities in loops, to represent how competing water use processes are operating, and to search for acceptable solutions through better coordination and collective scenario assessment.
>
> (Governance Author Team, 2010, p. 9)

The final project report concludes that the ComMod project contributed to improved communication and trust among multiple stakeholders. Examples of multiple stakeholders were the forest authorities and villagers in one of the Thailand sites and different groups of herders in one of the Bhutan sites. The ComMod project also helped to strengthen or establish local institutions to tackle natural resource management problems. Project activities led to: (1) a new regulation on irrigation water use in the Salaep catchment in Thailand; (2) the establishment of a watershed resource management committee in the Lingmuteychu watershed in Bhutan; (3) coordinated use of water tanks in Kengkhar village in Bhutan; and (4) a compromise between downstream shrimp farmers and upstream rice growers on the timing of saline water intake at an important sluice gate in a coastal site in Vietnam (Governance Author Team, 2010, pp. 10–11).

Summary of negotiation support activities of PN20: Sustaining Inclusive Collective Action that Links across Economic and Ecological Scales in Upper Watersheds, SCALES.

The objective of SCALES was to contribute to poverty alleviation in the upper watersheds of the tropics through improved collective action for watershed management within and across social-spatial scales. SCALES was implemented in one watershed in Kenya (Nyando) and two watersheds in Colombia (Coello and Fuquene). SCALES included conceptual modeling, participatory assessment of poverty and livelihood dynamics, participatory games that simulated watershed interactions, pilot development activities and *Conversatorio de Acción Ciudadana* (CAC). The CAC in the Coello watershed in Colombia was seen as a particular success. The main source of information about the CAC process used here is the impact assessment conducted by Córdoba et al. (2008).

The CAC method was developed by ASDES (Consultancies for Development Corporation), an NGO based in Cali, Colombia. ASDES had more than 20 years of experience in educating communities in citizen political action. It was first implemented with WWF as a mechanism for managing marine resources on the Colombia Pacific Coast. The CAC is consistent with the Constitution of Colombia, which enshrines the rights of citizens to hold their representatives accountable. In the CAC method, a community convenes a meeting with public and private institutions with the purpose of (a) solving social, political, economic, education or environmental problems, and (b) negotiating conflicts in relationships between the community and the state, the community and the territory, or between communities. Organizations that can help to resolve these problems are invited to the event through an official letter.

In the Coello watershed, the SCALES partners followed a three-step process toward the CAC: (1) sensitization; (2) preparation and implementation of the CAC; and (3) follow up. Loosely, the sensitization phase matches with the research category described as better understanding, the preparation phase was a decision-support activity, and the implementation and follow up can be described as negotiation support. To redress power imbalances in the Coello area, the project focused its better understanding and decision support activities on six community groups in the watershed. This helped those groups to reach decisions about the most important issues to raise with the other stakeholders in the formal meeting. The CAC process facilitated the exchange of information generated in multiple ways, including a field trip in which representatives of different agencies traveled together throughout the watershed.

The CAC process unfolded in 2007–8 and led to a total of 30 agreements between the communities and the other organizations involved. Several examples of real impact were already noted by the impact assessment team in December 2008 including: new opportunities for a local NGO to work across the watershed; greater interest and participation by provincial authorities in the management of protected areas; and mayors being more responsive to community concerns resulting in new community alarms and bridges. A total of US$665,000 was committed for projects agreed upon through the CAC process.

Participation in negotiation support

All three of the negotiation-support projects reviewed in this section took systematic approaches to the participation of key stakeholders in their research and boundary-spanning activities. Goodwill and a systematic approach, however, do not ensure the desired outcome. The PN50 report notes that many government agencies in the Mekong region were cautious about being involved in the types of multi-stakeholder platforms that the integrated watershed-management approach promotes. It is a reality that power in many countries is centralized and compartmentalized, making it very difficult to achieve the objectives of integrated water resource management. From the CPWF experience reviewed here, it seems that it may be much easier to achieve effective multi-stakeholder platforms at the local scale.

Accountability in negotiation support

All of the CPWF projects reviewed in this section had management and advisory group structures that attempted to increase accountability to stake-holder groups. All were implemented in different ways.

PN50 was implemented through M-POWER, which is a knowledge network of about 30 organizations. The partners in M-POWER are committed to continuing the network beyond the life span of any particular funded project such as PN50. M-POWER has annual Partners' Working Group meetings that provide opportunities for partners to learn from each other. They share experiences, synthesize results, develop new project ideas and jointly explore governance issues in the region. A logical multi-stakeholder forum that is involved with most of the same issues in the Mekong region is the Mekong River Commission, an inter-governmental created under the 1995 Mekong Agreement. The Ministers of Environment and Water of Cambodia, Laos, Thailand and Vietnam are Council members of the Mekong River Commission, with China and Myanmar as "Dialog Partners." It is encouraging to see that M-POWER is recognized as the Regional Panel of Experts of the Mekong River Commission's Basin Development Plan Program Phase 2. This formal recognition of the expertise of M-POWER shows the advantage of having PN50 implemented by a network of partners whose activities will continue well past the end date of the project.

Overall, PN20 was implemented through a consortium of international organizations, local universities, international NGOs and local NGOs. Case study sites were located in Kenya and Colombia. The project governance structure did not include a formal mechanism to ensure accountability from the local level to the overall project management. The project had its impacts in the specific sites, however, so it was more important that the implementing agencies were accountable to local stakeholders at each of the three sites. The accountability of the local NGOs involved in the Coello watershed were perhaps most important to the success of the CAC process. It will be those organizations that continue to interact with the local organizations after the project has concluded.

A similar situation may well have played out for the ComMod project. The project partners at the local level engaged with key stakeholders in each of the nine sites. Partners were trained in the relevant methods, then engaged to bring in local expertise and to define key problems.

Boundary objects in negotiation support

Of any field of knowledge-to-action, watershed management must be one of the easiest to develop boundary objects for. The illustration of the water cycle is a vivid way to teach people about the challenges of water allocation, water quality and watershed management. Of all CPWF projects, the ComMod project made most effective and systemic use of boundary objects in its work.

Conclusions and implications

The CPWF supported a diverse collection of research on water management institutions. It covered multiple aspects of both the IAD framework (Figure 6.2) and the CAPRI "box" (Figure 6.1) that considers institutions over time and space. As might be expected, however, most of institutional work in the CPWF is in the right hand, upper parts of the CAPRI box—thus involving more reliance on government and less on market mechanisms.

The vast majority of CPWF institutional analysis was multi-disciplinary. Papers were published in a wide range of journals, mostly multi-disciplinary, and were widely cited by other authors.

The CPWF has provided important insights into the governance of multi-scale social-ecological systems. It linked with major research initiatives and collaborative scholarly efforts in this area (Carl Folke and Elinor Ostrom, the Resilience Alliance). Analysis of the relationships between water management institutions and gender is a gap, with a few notable exceptions such as the work on multiple-use water services. The analytical frameworks used in this synthesis did not explicitly consider gender. We do not believe that this explains the lack of gendered findings since we deliberately looked for gender-related results in publications and other project documents. Partly because they are multi-scale, water resource management institutions are necessarily diverse, with no single solution good for all circumstances.

In decision support, salience and credibility were important but the latter did not exclusively come from the research outputs of the specific project. The most influential projects had good scientists and cutting-edge ideas but were not necessarily the best in terms of publications in the project. Credibility may come from the researchers as much as the research outputs. This is consistent with a strong history of engagement, well beyond the lifetime of an individual CPWF project, which in the specific locations appears to be very important for impact on institutions. In some cases this was through specific individuals who had established strong reputations in the local context. Sometimes it was through a long-term collaboration and sometimes through established local partners with good links to research and to the community.

The political context is extremely important. In some political contexts the idea of using multiple sources of knowledge, or of multiple agencies being involved in decision making, can be seen as threatening. In these circumstances, it may be easier to show impact through a direct knowledge-action pathway, but this may conflict with objectives related to inclusion and representation. For example, the M-POWER context was very challenging, while the CAC process in Colombia was supported by the national constitution. Scale also matters. It was much easier to show quick impact at a catchment scale (e.g., ComMod, CAC) than at the larger national or regional scale (e.g., M-POWER).

A general message is that we need to develop better ways to measure saliency and legitimacy of research so that "boundary work" can be assessed for what it is.

Notes

1   Because of time constraints, we did not review theses. A very large number of undergraduate and graduate theses were produced as part of CPWF from universities in the north and south.
2   Some papers were included in this analysis that were subsequently dropped from the review because they were not relevant to the review.
3   The product is on: netmap.wordpress.com/services-and-products/ (accessed 16 April 2014).

Appendices

Table 6.1 Number of publications on water management institutions, by project and year.

Project number	Pre 2007	2007	2008	2009	Post 2009	Total
10	3				2	5
17			1			1
19			1		1	2
20	1			3	2	6
22	1	1		1		3
25	2	2	1		2	7
28		1	1	2		4
40	2	2	1	1	1	7
42			3	1		4
46	1					1
47	2	3		2		7
50	6	3	2	9	1	21
Total	18	12	10	19	9	68

Table 6.2 Journals in which CPWF institutions' research was published.

Journal title	SCI subject area(s)	Impact factor
Agricultural Water Management	1, 10	1.998
Ambio	14	2.025
Asian Journal of Env and Disaster Mgnt	—	—
Colorado Journal of Int Env Law and Policy	—	—
Creighton Law Review	—	2.2
Current Opinion in Env Sustainability	14, 25	2.438
Dev. In Practice	25	1.03
Ecological Economics	—	2.713
Ecology and Society	14	3.31
Energy Efficiency	12	1.085
Env and Dev Economics	14	0.671
Field Methods	25	1.111
Forest Ecology and Management	1, 14	2.744
Gender, Technology and Development	14, 24, 25	—
Int. J. Water Resource Development	14	0.795
Int Journal of Sustainable Development	4	0.965
Irrigation and Drainage	1, 14	—
Journal of African Economies	11	0.574
Journal of Agricultural Education and Extension	1	2.62
Journal of Environment and Development	14	—
Journal of Land Use Science	10, 14, 25	—
J. World Assoc. Soil Water Conservation	14	—
Journal of Liberal Arts	25	—
Land Use Policy	1	2.292
Mountain Research and Development	14	0.676
Natural Resources Journal	14	—

Table 6.2 Continued

Journal title	SCI subject area(s)	Impact factor
Outlook on Agriculture	1	0.556
Physics & Chemistry of the Earth	10	1.11
Regional Environmental Change	10, 19	3.00
Review of Policy Research	25	0.646
Science as Culture	25	0.37
Simulation and Gaming	4	—
Water Alternatives	25	—
Water International	13, 14	1.145
Water Policy	14	0.648
Water Resources Management	13, 14	2.054
Water South Africa	14	0.911
World Development	11	1.537
World Ecology	14	—

Sources: SCI Subject Matter—SCImago Journal and Country Rank, http://www.scimagojr.com/journalrank.php. Impact factors: Journal websites. SCImago Journal and Country Rank. Subject areas: 1. Agricultural and Biological Sciences; 2. Arts and Humanities; 3. Biochemistry, Genetics and Molecular Biology; 4. Business, Management and Accounting; 5. Chemical Engineering; 6. Chemistry; 7. Computer Science; 8. Decision Sciences; 9. Dentistry; 10. Earth and Planetary Sciences; 11. Economics, Econometrics and Finance; 12. Energy; 13. Engineering; 14. Environmental Sciences; 15. Health Professions; 16. Immunology and Microbiology; 17. Materials Science; 18. Mathematics; 19. Medicine; 20. Multidisciplinary; 21. Neuroscience; 22. Nursing; 23. Pharmacology, Toxocology, Pharmacy; 24. Psychology; 25. Social Sciences; 26. Veterinary.

References

Baran, E., Jantunen, T. and Chheng, P. (2006) 'Developing a consultative Bayesian model for integrated management of aquatic resources: An inland coastal zone case study', in C. T. Hoanh, T. P. Tuong, J. W. Gowing, B. Hardy (eds), *Environment and livelihoods in tropical coastal zones: Managing agriculture–fishery–aquaculture conflicts,* CABI Publishing, UK, pp. 206–218.

Baran, E., Jantunen, T., Chheng, P. and Hoanh, C. T. (2010) 'Integrated management of aquatic resources: A Bayesian approach to water control and trade-offs in Southern Vietnam. Improving the productivity of the rice-shrimp system in the southwest coastal region of Bangladesh', in C. T. Hoanh, B. Szuster, S.-P. Kam, A. Noble and A. M. Ismail (eds), *Tropical deltas and coastal zones community, environment and food production at the land-water interface,* CABI Publishing, UK, pp. 133–143.

Barnaud, C., Promburom, P., Bousquet, F. and Trebuil, G. (2006) 'Companion modelling to facilitate collective land management by Akha villagers in upper northern Thailand', *Journal of the World Association for Soil and Water Conservation,* JI: 38–54.

Barnaud C., Promburom, T., Trébuil, G. and Bousquet, F. (2007) 'Evolving simulation and gaming to support collective watershed management in mountainous northern Thailand', *Simulation and Gaming,* vol. 38, pp. 398–420.

Barnaud, C., van Paassen, A., Trébuil, G., Promburom, T. and Bousquet, F. (2010) 'Dealing with power games in a companion modelling process: Lessons from community water management in Thailand Highlands', *Journal of Agricultural Education and Extension,* vol. 16, issue 1, pp. 55–74.

Becua, N., Neef, A., Schreinemachers, P. and Sangkapitux, C. (2008) 'Participatory computer simulation to support collective decision-making: Potential and limits of stakeholder involvement', *Land Use Policy*, vol. 25, no. 4, pp. 498–509.

Berger, T. and Schreinemachers, P. (2006) 'Creating agents and landscapes for multi-agent systems from random samples', *Ecology and Society*, vol. 11, no. 2, art. 19, ecologyandsociety.org/vol11/iss2/art19/ (accessed 17 April 2014).

Berger, T., Birner, R., McCarthy, N., Díaz, J. and Wittmer, H. (2007) 'Capturing the complexity of water uses and water users within a multi-agent framework', *Water Resources Management*, vol. 21, no. 1, pp. 129–148.

Bharati, L., Rodgers, C., Erdenberger, T., Plotnikova, M., Shumilov, S., Vlek, P. and Martin, N. (2008) 'Integration of economic and hydrologic models—An application to evaluate conjunctive irrigation water use strategies in the Volta Basin', *Agricultural Water Management*, vol. 95, no. 8, pp. 925–936.

Bousquet, F., Castella, J.-C., Trébuil, G., Barnaud, C., Boissau, S. and Kam, S.-P. (2007) 'Using multi-agent systems in a companion modelling approach for agroecosystem management in South-east Asia', *Outlook on Agriculture*, vol. 36, no. 1, pp. 57–62.

Buizer, J., Jacobs, K. and Cash, D. (2010) 'Making short-term climate forecasts useful: Linking science and action', *Proceedings of the National Academy of Sciences*, pnas.org/content/early/2012/01/18/0900518107.full.pdf+html?sid=1b8169f4-d36f-4b0b-beb2-f01cd4ba905d (accessed 17 April 2014).

Candelo, C., Cantillo, L., Gonzalez, J., Roldan, A. M. and Johnson, N. (2008) 'Empowering communities to co-manage natural resources: Impacts of the *Conversatorio de Acción Ciudadana*', *Proceedings of the CGIAR Challenge Program on Water and Food 2nd International Forum on Water and Food, Addis Ababa, Ethiopia, 10–14 November 2008*, CGIAR Challenge Program on Water and Food, Colombo, Sri Lanka.

Cardenas, J. C., Rodriguez, L. A. and Johnson, N. (2010) 'Collective action for watershed management: Field experiments in Colombia and Kenya', *Environment and Development Economics*, vol. 16, special issue 3, DOI: 10.1017/S1355770X 10000392.

Clark, W. C., Tomich, T. P., van Noordwijk, M., Guston, D., Catacutan, D., Dickson, N. M. and McNie, E. (2011) 'Boundary work for sustainable development: Natural resource management at the Consultative Group on International Agricultura Research (CGIAR)', *Proceedings of the National Academy of Science* (August 15, 2011), pnas.org/cgi/doi/10.1073/pnas.0900231108 (accessed 25 April 2014).

Córdoba, D., de León, C. and Douthwaite, B. (2008) *The Conversatorio of Citizen Action as a tool for generating collective action for integrated water management: Evaluation of the impact of the Project SCALES/PN20—the sustaining collective action linking economic and ecological scales in upper watersheds*, Final Report, Impact Assessment Project, CGIAR Challenge Program on Water and Food, Colombo, Sri Lanka.

CPWF (2013) *A conceptual framework to link collective action, scale and poverty*, CGIAR Challenge Program on Water and Food, Colombo, Sri Lanka.

Crow, B., Swallow, B. and Asamba, I. (2012) 'Community organized household water increases not only rural incomes, but also men's work', *World Development*, vol. 40, no. 3, pp. 528–541.

Di Gregorio, M., Hagedorn, K., Kirk, M., Korf, B., McCarthy, N., Meinzen-Dick, R. and Swallow, B. (2008) *Property rights, collective action, and poverty: The role of institutions for poverty reduction*, CAPRI Working Paper, no. 81. Washington DC, IFPRI, http://www.capri.cgiar.org/pdf/capriwp81.pdf (accessed 25 April 2014).

152 *Johnson, Swallow and Meinzen-Dick*

DWAF (2006) *Provision of water for small scale multiple use systems: A guide for municipalities*, Version 1, April 2006, Draft, Department of Water Affairs, Pretoria, South Africa.

Estrada, R. D., Quintero, M., Moreno, A. and Ravnborg, H. M. (2009) *Payment for environmental services as a mechanism for promoting rural development in the upper watersheds of the Tropics*, CPWF Project Report, PN22, CGIAR Challenge Program on Water and Food, Colombo, Sri Lanka.

Friend, R. M. and Blake, D. J. H. (2009) 'Negotiating trade-offs in water resources development in the Mekong Basin: Implications for fisheries and fishery-based livelihoods', *Water Policy*, vol. 11, no.1, pp. 13–30.

Fujimura, J. H. (1992) 'Crafting science: Standardized packages, boundary objects and translation', in A. Pickering (ed.) *Science as practice and culture*, University of Chicago Press, pp. 168–211.

Governance Author Team (2010) *Companion modeling for resilient water management: Stakeholders' perceptions of water dynamics and collective learning at the catchment scale*, GREEN Research Unit, Cirad, Montpellier, France.

Gowing, J. W., Tuong, T. P., Hoanh, C. T. and Khiem, N. T. (2006) 'Social and environmental impact of rapid change in the coastal zone of Vietnam: An assessment of sustainability issues', in C. T. Hoanh, T. P. Tuong, J. W. Gowing and B. Hardy (eds) *Environment and livelihoods in tropical coastal zones*, CABI Publishing, UK.

Gurung, T. R., Bousquet, F. and Trébuil, G. (2006) 'Companion modeling, conflict resolution, and institution building: Sharing irrigation water in the Lingmuteychu Watershed, Bhutan', *Ecology and Society*, vol. 11, no. 2, art. 36.

Guston, D. (1999) 'Stabilizing the boundary between politics and science: The role of the office of technology transfer as a boundary organization', *Social Studies of Science*, vol. 29, pp. 87–111.

Hagos, F., Haileslassie, A., Bekele, S., Mapedza, E. and Taffesse, T. (2011) 'Land and water institutions in the BNB: Setups and gaps for improved land and water management', *Review of Policy Research*, vol. 28, pp. 149–170.

Jack, B. K. (2009) 'Upstream–downstream transactions and watershed externalities: Experimental evidence from Kenya', *Ecological Economics*, vol. 68, no. 6, pp. 1813–1824.

Johnson, N., García, J., Rubiano, J. E., Quintero, M., Estrada, R. D., Mwangi, E., Peralta, A. and Granados, S. (2009) 'Water and poverty in two Colombian watersheds', *Water Alternatives*, vol. 2, no. 1, pp. 34–52.

Kanyoka, P., Farolfi S. and Morardet S. (2008) 'Households' preferences for multiple use water services in rural areas of South Africa: An analysis based on choice modelling', *Water SA*, vol. 34, no. 6, pp. 715—723 (IWRM Special Edition).

Khiem, N. T. and Hossain, M. (2010) 'Dynamics of livelihoods and resource use strategies in different ecosystems of the coastal zones of Bac Lieu', in C. T. Hoanh, B. Szuster, S-P. Kham, A. Noble and A. M. Ismail (eds) *Tropical deltas and coastal zones community, environment and food production at the land–water interface*, CABI Publishing, UK.

Knox, A., Meinzen-Dick, R. and Hazell, P. (1998) *Property rights, collective action and technologies for natural resource management*, CAPRI Working Paper 1, IFPRI, Washington DC, capri.cgiar.org/pdf/capriwp01.pdf (accessed 17 April 2014).

Lautze, J. and Giordano, M. (2005) 'Transboundary water law in Africa: Development, nature, and geography', *Natural Resources Journal*, vol. 45, no. 4, pp. 1053–1087.

Lautze, J. and Giordano, M. (2006) 'Equity in transboundary water law: Valuable paradigm or merely semantics? *Colorado Journal of International Environmental Law and Policy*, vol. 17, no. 1, pp. 89–122.

Lautze, J. and Giordano, M. (2007a) 'Demanding supply management and supplying demand management: Transboundary waters in Sub-Saharan Africa', *Journal of Environment and Development,* vol. 16, pp. 290–306.

Lautze, J. and Giordano, M. (2007b) 'A history of transboundary water law in Africa', in M. Grieco, M. Kitissou, and M. Ndulo (eds) *The hydropolitics of Africa: A contemporary challenge,* Cambridge Scholars Publishing, pp. 87–125.

Lebel, L., Garden, P. and Imamura, M. (2005) 'The politics of scale, position and place in the management of water resources in the Mekong region', *Ecology and Society,* vol. 10, no. 2, art. 18.

Lebel, L. (2006) 'Multi-level scenarios for exploring alternative futures for upper tributary watersheds in mainland Southeast Asia', *Mountain Research and Development,* vol. 26, pp. 263–273.

Lebel, L., Anderies, J. M., Campbell, B., Folke, C., Hatfield-Dodds, S., Hughes, T. P. and Wilson, J. (2006) 'Governance and the capacity to manage resilience in regional social-ecological systems', *Ecology and Society,* vol. 11, no. 1, art. 19.

Lebel, L. and Daniel, R. (2009) 'The governance of ecosystem services from tropical upland watersheds', *Current Opinion Environmental Sustainability,* vol. 1, pp. 61–68.

Lebel, L., Grothmann, T. and Siebenhüner, B. (2010) 'The role of social learning in adaptiveness: Insights from water management', *International Environmental Agreements: Politics, Law and Economics,* vol. 10, no. 4, pp. 333–353.

Lebel, P., Chaibu, P. and Lebel, L. (2009) 'Women farm fish: Gender and commercial fish cage culture on the Upper Ping River, northern Thailand', *Gender, Technology and Development,* vol. 13, pp. 199–224.

Lundy, M. (2004) 'Learning alliances with development partners: A framework for outscaling research results', in D. Pachico (ed.) *Scaling up and out: Achieving widespread impact through agricultural research,* Centro Internacional de Agricultura Tropical (CIAT), Cali, Colombia.

Ma, X., Xu, J. C. and Qian, J. (2008) 'Water resource management in a middle mountain watershed', *Mountain Research and Development,* vol. 28, no. 3/4, pp. 286–291.

Magombeyi, M. S., Rollin, D. and Lankford, B. (2008) 'The River Basin Game as a tool for collective water management at community level in South Africa', *Physics and Chemisty of the Earth,* vol. 33, pp. 873–880.

Manuta, J., Khrutmuang, S., Huaisai, D. and Lebel, L. (2006) 'Institutionalized incapacities and practice in flood disaster management in Thailand', *Science and Culture,* vol. 72, pp. 10–22.

Manzungu, E., Mpho, T. J. and Mpale-Mudanga, A. (2009) 'Continuing discontinuities: Local and state perspectives on cattle production and water management in Botswana', *Water Alternatives,* vol. 2, no. 2, pp. 205–224.

Merrey, D. J. and Sibanda, L. M. (2008) *Multiple Use Water Services (MUS) Project: Assessment of impacts and their pathways as a basis for learning lessons for future projects,* CPWF Impact Evaluation Report, CGIAR Challenge Program on Water and Food, Colombo, Sri Lanka.

Merrey, D. J. (2009) 'African models for transnational river basin organisations in Africa: an unexplored dimension', *Water Alternatives,* vol. 2 no. 2, pp. 183–204.

Molle, F. and Hoanh, C. T. (2009) *Implementing Integrated River Basin Management: Lessons from the Red River Basin, Vietnam,* IWMI Research Report No 131, Colombo, Sri Lanka.

Molle, F., Foran, T. and Kakonen, M. (eds) (2009) *Contested waterscapes in the Mekong region. Hydropower, livelihoods and governance,* Earthscan, 416pp.

North, D. C. (1990) *Institutions, institutional change and economic performance*, Cambridge University Press, 152pp.

Ostrom, E. (2005) *Understanding institutional diversity*, Princeton University Press, New Haven, CT, 384pp.

Penning de Vries, F. W. T. (2007) 'Learning Alliances for the broad implementation of an integrated approach to multiple sources, multiple uses and multiple users of water', *Water Resources Management*, vol. 21, pp. 79–95, musgroup.net/page/1055 (accessed 17 April 2014).

Poolman, M. and van de Giesen, N. (2006) 'Participation: rhetoric and reality—the importance of understanding stakeholders based on a case study in Upper East Ghana', *International Journal of Water Resources and Development*, vol. 22, no. 4, pp. 561–573.

Quintero, M., Wunder, S. and Estrada R. D. (2009) 'For services rendered? Modeling hydrology and livelihoods in Andean payments for environmental services schemes', *Forest Ecology and Management*, vol. 258, pp. 1871–1880.

Ruankaew, N., Le Page, C., Dumrongrojwattana, P., Barnaud, C., Gajaseni, N., van Paassen, A. and Trébuil, G. (2010) 'Companion modelling for integrated renewable resource management: A new collaborative approach to create common values for sustainable development', *International Journal of Sustainable Development and World Ecology*, vol. 17, no. 1, pp. 15–23.

Rubiano, J., Quintero, M., Estrada, R. D. and Moreno, A. (2006) 'Multi-scale analysis for promoting integrated watershed management', *Water International*, vol. 31, no. 3, 38pp.

Sajor, E. and Ongsakul, R. (2007) 'Mixed land use and equity in water governance in peri-urban Bangkok', *International Journal of Urban Regional Research*, vol. 31, no. 4, pp. 782–801.

Sajor, E. and Thu, M. (2009) 'Institutional and development issues in integrated water resource management of Saigon River', *Journal of Environment and Development*, vol. 18, no. 3, pp. 268–290.

Schiffer, E. and Peakes, J. (2009) 'An innovative approach to building stronger coalitions: the Net-Map Toolbox', *Development in Practice*, vol. 19, no. 1, pp. 103–105.

Schiffer, E. and Hauck, J. (2010) 'Net-Map: Collecting social network data and facilitating network learning through participatory influence network mapping', *Field Methods*, vol. 22, pp. 231–249.

Schreinemachers, P. and Berger, T. (2006) 'Land use decisions in developing countries and their representation in multi-agent systems', *Journal of Land Use Science*, vol. 1, no. 1, pp. 29–44.

Shah, T., Bhatt, S., Shah, R. K. and Talati, J. (2008) 'Groundwater governance through electricity supply management: Assessing an innovative intervention in Gujarat, western India', *Agricultural Water Management*, vol. 95, pp. 1233–1242.

Swallow, B. M., Johnson, N. L., Meinzen-Dick, R. S. and Knox, A. (2006) 'The challenges of inclusive cross-scale collective action in watersheds', *Water International*, vol. 30, no. 3, pp. 361–375.

Tsegaye, Y. B. E. and Berger, T. (2007) 'The agricultural technology: Market linkage under liberalization in Ghana: Evidence from micro data', *Journal of African Economies*, vol. 17, no. 1, pp. 62–84.

Uphoff, N. T. (1993) 'Creating institutional frameworks for self-managed local development', in P. Hall, R. de Guzman, C. M. Madduma Bandara and A. Kato (eds) *Multilateral cooperation for development in the twenty-first century: Training and research for regional development*, United Nations Centre for Regional Development, Nagoya.

van Koppen, B., Smits, S., Moriarty, P., Penning de Vries, F., Mikhail, M. and Boelee E. (2009) *Climbing the water ladder: Multiple-use water services for poverty reduction*, TP series, no. 52, IRC International Water and Sanitation Centre, The Hague and International Water Management Institute, Colombo, Sri Lanka, musgroup.net/home/activities/the_cpwf_mus_project/global_outputs/global_climbing_the_water _ladder_multiple_use_water_services_for_poverty_reduction/(language)/eng-GB (accessed 17 April 2014).

White, D. D., Wutich, A., Larson, K. L., Gober, P., Lant, T. and Senneville, C. (2010) 'Credibility, salience and legitimacy of boundary objects: Water managers' assessment of a simulation model in an immersive decision theatre', *Science and Public Policy*, vol. 37, pp. 219–232.

Xu, J. C., Yang, Y., Fox, J. and Yang, X. (2007) 'Forest transition, its causes and environmental consequences: Empirical evidence from Yunnan of Southwest China', *Tropical Ecology*, vol. 48, no. 2, pp. 137–150, tropecol.com/volumes/toc/en/toc48_2_e.htm (accessed 25 April 2014).

7 Partnerships, platforms and power

Amy Sullivan,[a*] *Terry Clayton,*[b]
Amanda Harding[c] and *Larry W. Harrington*[d]

[a]Food, Agriculture and Natural Resources Policy Analysis Network FANRPAN, Pretoria, South Africa; [b]CGIAR Challenge Program on Water and Food, Vientiane, Lao PDR; [c]CGIAR Challenge Program on Water and Food CPWF, Paris, France; [d]CGIAR Challenge Program on Water and Food CPWF, Ithaca, NY, USA; [*]Corresponding author, amysullivan3@gmail.com.

Introduction

Previous chapters described how the Challenge Program on Water and Food (CPWF) adopted a research-for-development (R4D) approach to address problems in complex adaptive water and food systems. In R4D, the entire research process, including outputs, can be the basis for strategic engagement with decision-makers. Engagement strategies feature the participation of development actors or boundary partners. They aim to modify decision-maker knowledge, attitudes and skills to influence policy and practice (outcomes). Engaging decision-makers from the outset allows them to contribute to define the problem, set priorities, and design and implement the research. Strategic engagement creates feedback loops to improve research itself. New information can improve researchers' and decision-makers' understanding of the problem in hand and how to address it (Figure 7.1).

In this chapter we further explore R4D engagement strategies and what these may look like in program implementation. We argue that innovations are embedded within a policy and political context and that engagement deals with power relationships. We found that partnerships and platforms[1] are central to successful engagement, but that these are also complicated by questions of power. We first explore the CPWF's experiences with partnerships and platforms and then discuss power issues in the context of R4D. We present two kinds of examples, from individual CPWF projects and from the CPWF itself as a reform program of the CGIAR.

Partnerships

Partnerships and networks are central to the CPWF's R4D approach. Basin and Project Leaders focused on developing partnerships and collaboration. The CPWF found that personal contacts and engaging in networks to create social

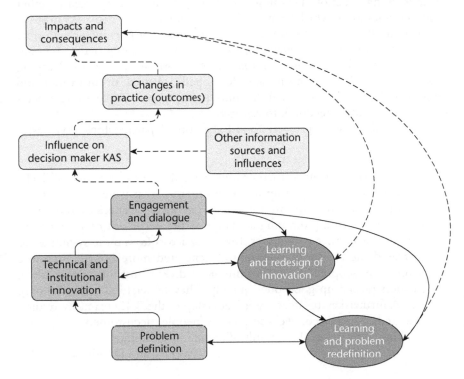

Figure 7.1 Conceptual framework showing where research, learning and engagement fit in the process of innovation. KAS = knowledge, attitudes and skills. Research outputs are more likely to change outcomes (decisions, practice) when they influence KAS.

Source: Harrington and van Brakel, Chapter 5 this volume.

capital[2] were important to advance R4D. Partnerships were important to accelerate the development of innovation (learning selection), and to engage with decision-makers to transform outputs to outcomes (theory of change). The CPWF also found that including development agencies and boundary partners in its teams facilitated taking innovation to scale, beyond the range of individuals' personal contacts.

Partnerships and learning selection

In Chapter 3, the authors posited that innovation was based on people trying novel ways to do things. If the attempt succeeds, they may decide to continue with the novelty or adapt it further. If conditions are not favorable to success they may abandon it. While they experiment, they interact with others who may influence what they decide to do with the novelty, which the authors called learning selection. Learning selection by many people linked together produces innovation, which provides the theoretical basis for networks in R4D.

Theoretically, networks can accelerate the process of learning selection by strengthening patterns of interaction among partners, and improving processes for assessing the fitness of novelties *in situ*. Networks encourage the learning process in which innovations are developed, adapted or discarded and emphasize the role of partners in the innovation process. The learning process focuses on providing feedback to research.

A study of Phase 1 projects found that broad partnerships gave faster learning.

> Projects [with expanded partnerships] contributed to resilience of liveli-hoods because they sped up learning processes that were cognizant and inclusive of different system scales. This provided the checks and balances necessary to avoid promoting a change to the detriment of a long-term trend, or of another system user. Involving actors from more system levels increased the ability to analyze and generated more benefit for more people. By scoping the environment of diverse institutions for ideas, partners picked up good ones quickly. They understood 'what is going on'. A further key to success was leadership of the R4D teams by results-oriented, committed, well-connected people, accustomed to systems thinking, which was also a result of broader partnerships.
>
> (Woolley and Douthwaite, 2011)

Partnerships and theories of change

Chapter 3 discussed the use of theory of change (ToC) in guiding engagement strategies. Quoting from Alvarez et al. (2010),

> A theory of change is the causal (or cause–effect) logic that links research activities to the desired changes in the [people] that a project or program is targeting to change. It describes the tactics and strategies, including working through partnerships and networks, thought necessary to achieve the desired changes in the target [people].

Consecutive iterations of a ToC provide information on who should be included as a partner for the purpose of engaging with which kinds of decision-makers. Outcome logic models make this explicit (Alvarez et al., 2010). ToC and outcome logic models were developed for each Phase 2 multi-project basin program.

The CPWF found that the range of partners needed to make progress in R4D depended on the issues or challenges being addressed. Moreover, different partners may be needed in different roles at different stages of R4D. For example, one set of partners may be helpful to define a challenge (for example the temporal and spatial incidence of drought and its effects on poverty). Another set of partners might develop drought-management strategies (drought-tolerant varieties, soil-cover strategies, market innovations for no-till farm equipment). Yet a different set of partners could assess the consequences (returns to investment in conservation agriculture, distribution of benefits across social groups) or engage with policymakers to explore policy alternatives to encourage drought-management practices. In study basins, the CPWF often represented just one piece of a larger agenda in which partner organizations had already invested. This led to questions and long discussions about attribution of results and outcomes.

Roles for CPWF in R4D

Key features of the CPWF's R4D were its roles as convener, negotiator, enabler, space provider and trusted broker (see Chapter 3). They promoted dialogue between key people in research, policy and development. With CPWF research outputs intended to contribute to development outcomes, partners brought quality research to the attention of policymakers and were able to engage them in its development.

R4D requires a mix of skills and partners with clear mandates, roles, perspectives and linkages, plus the ability or flexibility to work in partnerships toward longer-term goals. At the same time, it is important to maintain focus on research-based evidence for better-informed decision-making. The Basin Development Challenges (BDCs) of Phase 2 focused on local realities and people, took account of national policy frameworks and included relevant, knowledgeable partners across scales.

Partnerships in the CPWF

Partnerships and collaboration were central to the concept of the Challenge Programs (CPs) when they were proposed in 2001, and they continued to be so in the development of R4D in Phase 2 of the CPWF. Despite this, an informal survey in March–April 2013 indicated no common definition of partnerships within the CPWF. Individual Basin Leaders and research teams had their own definitions of partnerships, which guided their practice. This led the CPWF to ask, "Is it necessary to have a common definition of partnerships and collaboration within a program like the CPWF? Would that lead to more effective partnerships, or is partnership entirely a matter of individuals in context?" The literature suggested that while context is an important driver, there are underlying commonalities and best practices.

One way to distinguish different kinds of partnerships is by the functions they perform. Some partners helped the CPWF understand development challenges. Some partners worked with it to develop strategies to address these challenges. In some cases the CPWF was just one of several partners in a platform of which it was not the convener. In other cases, partners helped with engagement and negotiations where there were political, social and other sensitivities.

Sometimes the best partnerships were those where partners co-invested. Examples are NGOs and the private sector in goat auction sites in Zimbabwe, hydropower companies in constructed wetlands in Laos, and the Ministry of Environment in benefit-sharing mechanisms in Peru.

Platforms

The CPWF used several types of platforms, all focused on engagement, including innovation platforms and multi-stakeholder platforms. In general, an engagement platform is a space where individuals and organizations with different backgrounds and interests can assemble to diagnose problems, identify opportunities and implement solutions.

Engagement platforms redefine who we think of as decision-makers. Decision-makers are normally assumed to be senior government officials. We say assumed because despite the many references to decision-makers in the literature, they are seldom identified. Adequate design of ToCs helps identify decision-makers previously 'hidden' from development processes. By identifying and including these under-represented people and positions, engagement platforms help generate meaningful contextualized interaction among various actors with power and mandates to address real issues.

Within the context of the engagement platform, however, everyone can be a decision-maker. Decisions are made by the group, which may include senior government officials, business owners, trade unionists, NGO staff, members of farmers' associations and scientists. Other members of the group may not consider that scientific evidence is superior to any other kind of evidence. The research may point in one direction, but the group may have valid reasons for going in a different direction.

"Innovation platforms" is the term used in the CPWF for very specific engagement platforms in the three African basins, where they were used to link researchers, end users and boundary partners with a range of technologies (van Rooyen and Homann-Kee Tui, 2009; Duncan, 2012; Cullen, 2013). In contrast, multi-stakeholder platforms are where people from different groups with different interests meet to discuss contentious issues. The platform offers them a neutral and safe environment in which they can discuss the issues, and was a feature of the Mekong BDC (Geheb, 2012).

Assertions

The challenges facing water and food management in basins are complex and require different types of platforms that bring together people with multiple perspectives. Over its 11 years, the CPWF invested in engagement platforms across a wide range of scales to address a wide range of challenges (Clayton, 2013).

The most successful engagement platforms are demand driven, solution oriented, evolve over time and embrace multiple perspectives

An engagement platform is not a committee but a dynamic entity. Whilst there is a core group, members come and go as the problems change. Different stakeholders can be involved at different times, depending on the issues being discussed. The challenge that the platform facilitators face is to define and engage those who should be involved at critical junctures as not everyone need be involved all the time. At different stages different people might be involved.

Build on what is already there rather than set up new platforms and systems

It takes time and other resources to set up platforms, to get members to understand what platforms are and how they function, to build trust and develop a collective vision and agenda. Inviting multiple perspectives also means that there is a need to understand different agendas and sometimes conflicting mandates. For all these reasons, engagement platforms are best developed around existing relationships, networks and structures. In the Limpopo Basin, the CPWF used existing national-level platforms (through the Food, Agriculture and Natural Resources Policy Analysis Network (FANRPAN), the Global Water Partnership Southern Africa and WaterNet, while also linking to the Southern African Development Community (SADC)) and the Limpopo Watercourse Commission to channel its research results. This ensured that results related to regionally relevant development priorities and that the information decision-makers requested came from the research.

Engagement platforms are not neutral mechanisms: They aim to promote change so they are disruptive by nature

Changing existing dynamics is likely to distribute consequences across groups and may have unanticipated results. Engagement platforms can help balance vested interests in policy-making processes but unless power dynamics are recognized and addressed, engagement platforms can reinforce existing inequalities. Addressing power and representation during the stage of setting up a project can help make engagement platforms more equitable and effective.

Engagement platforms are based on assumptions that members represent the various groups involved and are able to work together for their mutual benefit. They assume that better communication and knowledge sharing will help people understand each other's perspectives. They further assume that people can identify and agree on a common problem and work together to solve it. These assumptions need to be challenged in each case.

Engagement platforms can be useful vehicles to explore strategies to boost productivity, improve natural resources management, strengthen value chains and adapt to climate change

Engagement platforms can address a single issue or complex problems involving a wide range of people within farming communities. Individual stakeholders need the right reasons and incentives to make a platform work. Often this involves market incentives and removing barriers to benefits, which may directly challenge the status quo. Excellent facilitation is needed to turn engagement platforms into win–win endeavors, including acknowledging short- versus long-term engagement and outcomes.

Well-facilitated engagement platforms link different sectors and levels, which stimulates horizontal and vertical coordination for greater impact

Horizontal links refer to collaboration between platforms at the same institutional level to strengthen their bargaining position or for learning. Collaboration does not necessarily mean establishing multiple platforms. Usually it is better to link a single engagement platform to other organizations, networks or individuals. Engagement platforms may allow participants some freedom to explore issues that are high risk or controversial at a national level. In diagnosing issues and designing and implementing solutions, engagement platforms often evolve toward vertical integration.

Platforms can empower local actors to hold authority to account

Giving stakeholders a voice is one of the prime functions of an engagement platform. Local platforms connect local actors with actors who perform other functions. Research findings and solutions at those local levels may influence the role of actors performing at higher levels and may help source evidence that points to responsibilities of these higher-level actors.

Markets provide clear incentives for investments in production

Engagement platforms may help reduce the cost of searching for and reaching markets. They also allow people who understand the challenges and opportunities in the local system to devise and test solutions. Engagement platforms are tools for pooling knowledge across the agricultural business, education, research and extension systems. They generate, disseminate and use engagement to reduce transaction costs. Engagement platforms can target markets in, for example, school feeding programs, military institutions, hospitals, supermarkets, processors and commodity exchanges.

Power

Power and poverty

The original objective of the CPWF was to increase water productivity as a means of increasing food production and decreasing malnourishment and rural poverty. The focus was to be "in river basins with low average incomes and high physical, economic or environmental water scarcity or water stress, with a specific focus on low-income groups within these areas" (see Chapter 1; CPWF Consortium, 2002, p. 1). As the Program moved towards an explicit R4D approach, it maintained a focus on development and rural poverty.

Chapter 1 discussed the concept of poverty in terms of social exclusion and power: "[M]easures of poverty [range] from head counts of people living on a certain minimum amount of income to people-centered approaches of how well people meet their livelihood goals . . . [The CPWF] also included the concept of social exclusion acknowledging that multiple forms of discrimination impact severely on the poor and their capacity to influence decisions that directly affect their lives . . . Human agency is what poor people can do for themselves, and empowerment is creating conditions that allow them to do so." The notion of poverty as a lack of freedom emphasizes the impact of the institutions (organizations but also norms and frameworks of behavior) on decision-making as individuals and households pursue their livelihoods (Kemp-Benedict et al., 2011).

CPWF research in the Basin Focal Projects identified five "aspects of water-related poverty:

- "*Scarcity:* where people are challenged to meet their livelihood goals as a result of water scarcity.
- *Lack of access:* where people [from some ethnic or social groups] lack equitable access to water.
- *Low productivity:* where people acquire insufficient benefit from water use.
- *Chronic vulnerability:* where people are vulnerable to relatively predictable and repeated water-related hazards such as seasonal floods and droughts, or endemic disease.
- *Acute vulnerability:* where people suffer an impaired ability to achieve livelihood goals as a consequence of large, irregular and episodic water-related hazards" (Kemp-Benedict et al., 2011).

These aspects of water-related poverty are interconnected among themselves and with agency[3] and empowerment. Power relationships are fundamental to access to water and other resources while access itself is one of several dimensions of water scarcity (see Chapter 2). The productivity of water and other resources is influenced by access to inputs, credit and product markets but the poor often have limited access to these markets, goods and services. Vulnerability to water-related hazards increases risk to the point where the poor often cannot invest in practices that raise productivity (Scoones, 1996).

A related aspect of poverty, not included in the list above, is externalities. Powerful people upstream, through their land and water management practices, can impose costs on less-powerful downstream communities. The costs may be increased water scarcity, lack of access to resources or increased vulnerability. In the Andean basins, for example, grazing of fragile alpine wetlands by livestock can reduce downstream water availability by reducing water capture and increasing siltation. Conversely, powerful people downstream may use political or financial leverage to reduce the access of upstream communities to land and water to prevent negative externalities. For example, downstream communities might persuade policymakers to name upstream regions as no-access conservation reserves.

There are several examples from CPWF projects where combinations of power, resource access, vulnerability, externalities and poverty were linked.

In the Ganges Basin, community-based fisheries improved livelihoods in seasonally flooded areas in Bangladesh. Even the landless poor benefitted when they were allowed unlimited fishing by hooks and lines, but not nets (PN35)[4] (Sheriff et al., 2010). Sustainability of community-based fisheries and expanding them to new areas, however, depends on the ability of communities to maintain lease rights to flooded areas. In this they compete with private investors whose power relationships influence who gains access to the leases. The poor can easily be excluded (Toufique and Gregory, 2008; Collis et al., 2011).

In the Mekong Basin, CPWF projects worked on the nexus between hydropower, food and poverty. They found that hydropower dams, which benefit urban and industrial centers, often impose costs on the rural poor living downstream from the dams. Regarding flood and disaster management:

> [P]romises of protection are often made in earth or concrete: upstream dams . . . will regulate river flows; diversions will take the water around and past the city; [longer and higher] dykes . . . will hold back the flood waters; drains, pumps and tunnels will move water out faster. Flood management policies, measures and practices in the greater Mekong region, intended to reduce risks, however, frequently shift risks [on to] already vulnerable and disadvantaged groups. Promises of protection and how they are pursued can be explained in terms of beliefs, interests, and power.
>
> (PN50)[4] (Lebel et al., 2010)

Hydropower dams on the Mekong River are forecast to reduce the productivity of fisheries, critical to the livelihoods of millions of poor people living downstream. "There are good examples and verifiable science from around the world to indicate that dams have a significant negative impact on fisheries, in some cases driving them to collapse. The degree of impact will vary and depends on dam location, river hydrodynamics, and dam management" (Pukinskis and Geheb, 2012). In the Mekong Basin, the negative externalities of upstream dams could be catastrophic for Cambodia if they compromise the seasonal ebb-and-flow of the Tonle Sap and its fishery. Although the livelihoods of large numbers of poor families living downstream from dam sites depend on fisheries (Kam, 2010), they have little voice in the dialogue on the consequences of dams on the fisheries (MK5)[4] (Pukinskis and Geheb, 2012).

In the Andes basins, poor land and water management by subsistence farmers in the highlands lowered water quality and gave less-reliable dry-season flows downstream. The negative externality affected wealthier downstream users—urban communities, hydropower companies, recreational users and commercial irrigated farms. CPWF projects helped institute mechanisms for sharing benefits and costs in which upstream communities were compensated for using practices that generate positive, not negative, externalities. In some Andean basins, downstream water users chose to negotiate strategies that were equitable and provided positive ecosystem services (PN20, AN1-3)[4] (Escobar, 2012; Quintero, 2012; Saravia, 2012).

Power, gender, and the distribution of costs and benefits of innovation

In complex rural livelihoods, innovation often does not benefit everyone. Some groups may benefit from an innovation while other groups receive no benefit or are harmed by it. The least powerful are often those who are harmed. Problems with the distribution of the costs and benefits of innovation across groups may take many forms: upstream versus downstream water users; land owners versus landless; hydropower operators versus fishers; irrigated farmers versus pastoralists; youth versus age, and so on. In this section, however, we focus on gender bias as this is at the forefront of a more general problem of equitable distribution across social groups of the costs and benefits of innovation.

Addressing gender inequities required multiple approaches that recognized that attitudes to gender are complex social constructs within project teams, rural communities and institutions at all levels. Women's concerns were excluded a lot, often unconsciously, in designing and implementing innovations. By understanding the many forms of gender bias, the CPWF R4D community integrated concerns of power and voice, which the poorest and most vulnerable people often lack. R4D had to be mindful of gender questions so that women could access resources and engage in the development process. We needed to examine attitudes to gender within institutions so that researchers and implementers understood their role in addressing gender inequalities.

Research on water productivity of livestock in Phase 1 confirmed that the design, timing and labor requirements of some technologies affected men and women differently. Some technologies to increase livestock water productivity (LWP) made more work for women but gave them fewer benefits. Men and women benefitted differently from improved LWP and especially the type of livestock targeted. Smaller livestock largely benefitted women, and in so doing improved the education and health of children in poorer households. Improving LWP of one class of animal, however, does not always reduce poverty of the whole household or community. If increased LWP is to improve the livelihoods of both men and women, the interventions must consider the gender and power relations of the community (PN28)[4] (Mapezda et al., 2008).

Soil erosion and unrestricted livestock grazing cause feed shortage and soil degradation erosion in the Nile Basin in Ethiopia. A CPWF project developed different management practices for mixed crop–livestock systems. One of them was to enclose cattle in fenced fields rather than allow unrestricted open grazing on common land. The proposal had unexpected negative consequences for some social groups, including women.

The project overlooked that the community opposed enclosing communal grazing land, which is:

> [A]n open space accessible by the households living around it. [The common] is used for a variety of community gatherings . . . which are important in the maintenance of key social networks . . . Communal grazing areas are particularly important for households without livestock who rely on these areas for dung collection . . . which makes a vital contribution to local livelihoods. Enclosing grazing areas and keeping livestock at home denies vulnerable members of the community access to this resource.
>
> (Cullen, 2013)

"Women [were also concerned] that these changes could [affect] their children's safety [because in] rural areas of Ethiopia . . . children [tend the] livestock." If the livestock were to be enclosed, it would be more difficult for mothers to follow the movements of their children. "They were therefore reluctant to engage with the proposed interventions" (Nile 2) (Cullen, 2013). The problem remains unresolved.

In the Volta Basin, introducing small-scale irrigation changed the distribution of costs and benefits between irrigated farmers and pastoralists:

> [T]he main purpose of small reservoirs [changed] from livestock watering to small scale irrigation [using] treadle pumps, motor pumps, and drip irrigation. In Burkina Faso . . . small scale irrigation . . . increased the cropping area and changed the landscape around small reservoirs and wells. This made livestock management difficult in areas with strong pastoralist traditions like the Sahel.
>
> (Douxchamps et al., 2012b, 2012a)

To avoid damaging crops while still watering their herds, pastoralists had to avoid the cropping areas and cover longer distances to reach the grazing lands (Volta 1).

Power, mandates and legitimacy

In R4D, outputs from research should be anticipated by strategic partners and decision-makers and meet expressed needs. Engagement strategies aim to modify decision-makers' knowledge, attitudes and skills and so influence policy and practice, providing outcomes. Outputs can be used by individual decision-makers to understand a problem better and devise better solutions. When multiple decision-makers seek to engage in dialogue and negotiation to deal with contentious issues, outputs can be used to support the negotiations (see Chapter 6).

Power relationships, however, influence the course of dialogue and negotiation and thus the process of moving from outputs to outcomes. R4D practitioners must take account of mandates (who is responsible for what) by including individuals and institutions with relevant mandates in the engagement process. This is not easy. First, we must define relevance in terms of relevant to whom or to what. R4D practitioners must be mindful that many institutions have multiple and sometimes conflicting mandates and that several institutions may have overlapping mandates.

In the Limpopo Basin, for example, the Limpopo BDC operated with an acute awareness of mandates. The basin program sought to link its R4D with on-going political processes such as policies and priorities set by the Common Market for Eastern and Southern Africa, the Comprehensive Africa Agriculture Development Program, the SADC and basin commissions. This engagement helped link local innovation processes to higher-level processes and initiatives.

In Andean basins in Colombia, national policy required government entities to engage in negotiations with communities on contentious issues of resource management. Regional autonomous environmental authorities were mandated to facilitate negotiations. CPWF partners worked with communities to strengthen their capacity to engage in these negotiations, which resulted in binding agreements between communities and public sector agencies (PN20)[4] (Candelo et al., 2008; Johnson, 2009).

CPWF projects improved water control in polders in coastal Bangladesh, which enabled farmers to grow three crops each year instead of only one. Improved water control, however, involved local engineering departments (LEDs), which control infrastructure within polders, and the national Water Development Board, which controls polder development and maintenance. The national Planning Commission prioritizes and coordinates the activities of government entities. CPWF projects included LEDs, the Water Development Board and the Planning Commission in their engagement strategies for water control in polders (Ganges 3 and Ganges 5)[4] (George and Meisner, 2013).

A related issue is how to deal with accountability. Often entities with the power to make decisions are not accountable to the groups who are affected by their decisions. Worse, they sometimes also lack functional accountability. Moreover, powerful individuals or groups can shift the agenda in their favor over time, contributing further to the exclusion of the poor and vulnerable.

R4D practitioners must be aware of the potential for the elite to capture innovation platforms and engagement processes, and work to prevent it. They must also be accountable to stakeholders and partners, encouraging active feedback and transparency.

Looking inwards

In the above sections, we discussed power, partnerships and platforms in the context of the CPWF experience in R4D in basins. In this section, we discuss these issues in the context of the CPWF as a reform program within the CGIAR.

As the preceding chapters make clear, the CPWF only gradually came to portray its work in terms of R4D. R4D became a tool for doing research and a narrative for explaining success as basin teams turned more toward this new way of doing business. Ironically, there were no well-established pathways for delivering these lessons to the emerging CGIAR Research Programs (CRPs) within the CGIAR Centers engaged in reform.

Among the chapters of this book only Chapter 4, the institutional history, acknowledges the struggle to convince stakeholders within the CGIAR system that the CPWF R4D approach merited serious consideration. Clearly, the CPWF did not reach its goal of contributing to larger CGIAR-wide reform. To understand better the demands R4D will put on institutions, their individuals, partnerships and networks, we must look inward.

The CGIAR launched the CPWF as an experimental program with a new governance structure and business plan and with a "new quality of partner-ships" so as to "[change] the way [it did] business." Within the CPWF, the implicit assumption was that the CGIAR would learn from the results of the CPWF experiment and mainstream its more successful elements. In retrospect the assumption was naïve.

Within every organization there is tension between conservatism and change. Conservatism preserves the status quo and is based on the usually good evidence that "business as usual has worked for us so far." Adapting to change is always a risk and more often ends in failure than success (Ormerod, 2005). Moreover, there are real limits to organization or institutional change without an overhaul of mandates, personnel and incentive structures.

The CGIAR is not the only organization with a history of initiating reform processes then abandoning them for the next reform. The Challenge Programs were one of several such waves of reform in the CGIAR, each accompanied by remarkably similar discourses (Box 7.1).

Before the Challenge programs were the Ecoregional Programs initiated by a Technical Advisory Committee Working Group in the early 1990s (TAC Secretariat, 1993). This is shown as the Expansion period in Figure 7.2 and the short-lived integrated natural resource management initiative (CGIAR, 2000) in Figure 7.3. A decade after the Challenge Programs a new reform initiative was launched: the CRPs.

Box 7.1 The discourse of change in the CGIAR remains remarkably stable

- *Integrated natural resource management (INRM)*: "The shift from empirical to process-oriented research and use of system approaches will allow us to focus on ways of making ecosystems and natural resources managers such as farmers and others more capable of adapting positively in response to change, to work at multiple scales, and to suggest ways of dealing with the tradeoffs that are inevitable in various resource management options" (CGIAR, 2000).
- *Ecoregional approach*: "For the CGIAR to meet these challenges, changes in programming and organization that respond both to these changing global circumstances and to past operation inefficiencies are demanded. The second dimension is to adopt a new spirit of partnership with other research groups, and to design new mechanisms for closer integration across country, regional and international levels to bring greater coherence and efficiency to the global agricultural research system as a whole" (TAC, 1993).
- *Global Challenge Programs*: "The [CGIAR] System focuses the major part of its efforts on large multi-institutional research programs, which address specific problem areas using the expertise and competence of existing and new Center programs, and other partners. Most research programs are identified through a process which pulls in the suggestions of on-the-ground partners and potential new allies. The approach to problems is defined within an overarching vision of how the best science, together with other knowledge, can address the most urgent issues in a manner which will reduce poverty and promote development" (CGIAR, 2001).
- *CGIAR Research Programs*: "It has been recognized for more than a decade that the ever more complex issues facing agricultural research for development require an innovative approach to research. No single research institution working alone can address the critically important issues of global climate change, agriculture, and food security and rural poverty" (CGIAR, undated b).

The history of the CGIAR represents an orderly decadal progression as in Figure 7.2 and summarizes that progression as: "Sweeping changes in the first decade of the 21st century transformed our loose coalition into a streamlined global partnership working as one" (CGIAR undated a). An earlier depiction (Figure 7.3) shows the same evolution as a progression of topics as seen by the CGIAR Centers.

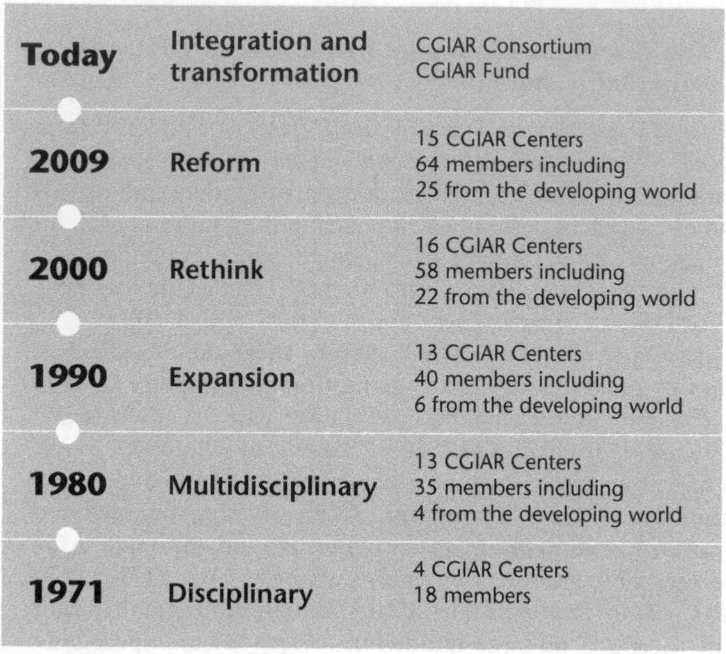

Today
Integration and transformation
CGIAR Consortium
CGIAR Fund
2009
Reform
15 CGIAR Centers
64 members including
25 from the developing world
2000
Rethink
16 CGIAR Centers
58 members including
22 from the developing world
1990
Expansion
13 CGIAR Centers
40 members including
6 from the developing world
1980
Multidisciplinary
13 CGIAR Centers
35 members including
4 from the developing world
1971
Disciplinary
4 CGIAR Centers
18 members

Figure 7.2 Institutional history as an orderly progression.
Source: CGIAR, undated a.

2005
Sustaining Biodiversity
Genetic Improvement
Agricultural Diversification
Natural Resource Management
Policies/ Institutions
2005
1990
Agroforestry and Forestry
Natural Resource Management
Living Aquatic Resources
1990
1980
Food Policy
Institutional Strengthening
1980
Mid 1970s
Genetic Resources
Dry Areas
West African Rice
Mid 1970s
Early 1970s
Tropics and Semi-Arid Tropics
Roots and Tubers
Livestock and Pastures
Cereals
Legumes
Early 1970s
1960s
Rice
Wheat
Maize
1960s

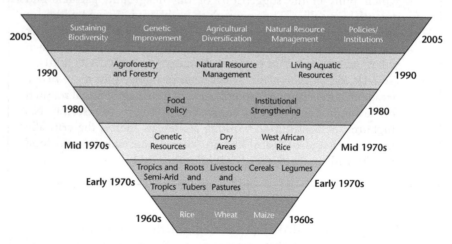

Figure 7.3 Institutional history as a progression of topics.
Source: Özgediz 2012, p. 129.

Change management has become a popular topic with a large academic and popular literature, but little consensus. Nevertheless, individuals or groups who wish to pursue R4D in their own organizations should start by looking for the fault lines within them. Who is predisposed to change? What skill sets are needed to help get research from outputs to outcomes? Simply passing the baton (outputs) to another partner to market to decision-makers will not work. The outputs themselves need to be planned and developed with the ultimate goal of feeding into specific decision-making processes.

The CPWF came to understand engagement and innovation platforms to convey that understanding to its CGIAR stakeholders and emerging CRPs. But the CPWF had difficulties in adapting its strategy to the quick emergence and rapid evolution of CRPs, not knowing exactly where and how those were developing.

Reassess incentives for learning

Any organization that wants to pursue R4D must reassess incentives for learning and engaging toward this end. Mbabu and Hall (2012) maintain that the learning system is the main tool for R4D. They emphasize the importance of a research culture that supports institutional learning and a change agenda.

Throughout this book authors have made repeated references to the importance of learning. The authors of Chapter 6 applied an institutional analysis-for-development framework to assess what CPWF researchers learned about action arenas and institutional processes. Subsequently they applied a boundary framework to examine how research on institutions influenced stakeholders and decision-making processes. Many practical lessons emerged that could be tested in other contexts and at other scales.

Whether or not and where the lessons learned are ever tested depends very much on how learning is defined and rewarded by organizations. Where researchers' performance is assessed mainly in terms of the number of publications and their citation index, learning for learning's sake is rewarded. In such a system, researchers and peers decide the worthiness and quality of research, mainly for each other. This could be characterized as research for research rather than R4D.

Despite what was set out in basin or project outcome pathways there was little incentive for researchers to step off the learning treadmill to engage for outcomes on the ground. This seems an inherent contradiction in using traditional research institutions to implement R4D for impact.

Within the organizational cultures of most mainstream research institutions, researchers receive tangible rewards for their publication productivity in the form of salary increases and promotion. Publications also enhance the reputations of individual researchers, making them more employable in the future. Within the CPWF, rewards for achieving outcomes were largely intrinsic. Because many of the CPWF partners were professionals in academic or research institutions, they simultaneously pursued scholarly publications by including them as outputs in their project proposals.

If outcomes are truly sought, one approach would be to build incentives into partnership contracts. A small but important step would be to make final payment contingent on distribution of outputs to boundary partners. For example, a CPWF Mekong Phase 1 project produced a *Guidance Manual on Agroecosystems Assessment in Cambodia*. It was launched by an inexpensive communications campaign that distributed 430 of the 500 printed copies to 63 individuals and institutions in 15 countries within three months. It did not produce big outcomes in terms of changes to government policy or new university curricula. But the chances of big outcomes are much increased with the small outcome of distributing the pamphlet.

In the Limpopo Basin a group of key partners and strategic stakeholders participated in a "science roll out" meeting with Limpopo BDC researchers. Participants shared and discussed outputs (decision-support tools, interactive databases on small reservoirs, rehabilitation guidelines, etc.). The event set the stage for deeper communication and collaboration between research partners and implementing agencies around specific outputs. It will continue beyond the lifetime of the CPWF. The process of getting from outputs to outcomes was facilitated through existing networks and partnerships, which had been nurtured from the beginning of Phase 2 of CPWF.

Organizational incentives

Organizations also have incentives, which ultimately boil down to funding, which in turn is dependent on how donors perceive the relevance, credibility and salience of the organization (see Chapter 6). This is why the CGIAR started the Global Challenge Programs, "to change the way [it did] business." But the Challenge Programs created an immediate paradox for their host Centers. If the challenge programs were allowed too much independence— over what direction they took, who they engaged with and what aims they pursued—the host Center still remained legally responsible for those actions. To reduce risk, Center Boards demanded tight central control over Challenge Program activities. (See Chapter 4 for more detail.)

For others attempting change, we suggest that changing the incentives for individual researchers could be a lever for institutional change. Willingness to make even small changes in the incentive structure could signal commitment to change. But it needs to be installed at the very beginning of the change process. Early in the formulation of the CRPs, Woolley and colleagues offered a critical analysis of management and financial arrangements that provided good guidelines (Woolley et al., 2009).

Understand your philosophical underpinnings

It is important to remember that the CPWF was not conceived as an R4D project. It was planned as a "new quality of partnership" with a non-Center-dominated governance structure aimed at achieving the accepted objective of "more crop per drop." Even with this acceptable aim, there was resistance to the new Challenge Programs. First, the Challenge Programs were perceived as competing with the CGIAR Centers for funding. Despite this being shown to be false, the perception continued to haunt the CPWF. Then there was resistance to changing theme-oriented research to research based on BDCs, although the change was in response to the criticism that the CPWF lacked focus. Finally, there was the on-going constraint of fiducial dependence on the host Center and limited managerial autonomy. Institutes set boundaries on who defines the problems and what sort of problems they may define.

The predominant model in public agricultural research institutes remains one of improving water productivity to increase the food supply to feed 9.6 billion people by 2050. Alternative views are not encouraged; for example, the alternative view that there is currently enough food for everyone and the problems are waste, poor distribution and self-interest policies (Moore-Lappe et al., 1986; Thurow and Kilman, 2009).

An example is that the CPWF questioned the orthodoxy that water scarcity is the defining crisis of the 21st century and that the solution is to increase water productivity. Working with partners in ten basins, the CPWF found that water scarcity is not strongly correlated with poverty. In doing so, CPWF redefined addressing water scarcity as a means of achieving broader development goals, including poverty reduction, rather than an end in itself (Chapter 4). This was regarded by some as an unorthodox view, but it pales in comparison to the far more serious break with conformity that R4D represents to some others.

Shift or schism

Throughout this book, we have described the evolution of the CPWF as "a shift in our understanding of developmental processes and the role of research." Further consideration suggests that the shift was a schism between the assumptions of theories of modernization and of participation.

Much development research is implicitly aligned with modernization theory. Modernization theory emerged from post-Second World War concerns of economists and policymakers, mainly in the USA, about unrest in newly emancipated nations and the threat of Soviet expansionism. Theorists created a category of nations they called "third world" or "developing," and diagnosed the underlying causes for why they were not yet developed. They proposed a simple remedy: changes in ideas will transform their behavior (Waisbord, 2000).

Development researchers then embarked on a mission for several decades to transmit information on modern values to modify behaviors according to development objectives, mainly to maximize profit for economic growth. Researchers insisted that it was their duty to influence policymakers by providing them with the information they lacked—information best obtained, of course, from research. When policymakers were not influenced, researchers may have conceded that their information was not packaged appropriately. But they seldom challenged the underlying assumption that policymakers lack information (Waisbord, 2001).

In contrast, R4D does not consider policymakers as mere recipients of information generated through research. It treats stakeholders in general and policymakers in particular as principal participants in a dynamic social process of innovation. The process is not controlled by researchers and research findings are only one of several sources of information. In Chapter 1 (see the section Poverty and development: The broader context) the authors wrote about the CPWF that "People-centered perspectives allowed the CPWF to consider the causes of poverty including the importance of human agency, empowerment, and institutional accountability. Human agency is what poor people can do for themselves, and empowerment is creating conditions that allow them to do so. These perspectives not only recognize the strategic importance of economic development, but the role of institutions as possible root causes of poverty."

Chapter 3 listed principles of R4D that emerged from practice within the CPWF, principles that reflect a philosophy sometimes at odds with the philosophy of mainstream development research:

* A focus on generating innovation amongst stakeholders;
* Embedding innovation processes in local institutional, policy and political contexts;
* Involvement and continuous interaction with stakeholders; and
* Science and research as an activity distributed throughout the economy and involving a dense network of people.

The assumptions of participatory theory underlie these statements. Participatory theory was a critical reaction against modernization theory, yet both theories see knowledge (research) as a means of influence and a lever for change. An important difference is the nature of that knowledge. Mainstream research sees knowledge as fundamental truth. R4D sees knowledge as socially constructed. Each leads to very different theories of change. In R4D, "engagement" is central to change and is manifested in partnerships and platforms.

In summary, the experience of the CPWF in R4D explains how and why the CGIAR struggles to implement innovative, partnership-based, cross-disciplinary, cross-scale, outcome-oriented research. Successful change is demand driven and is disruptive, not neutral. It empowers partners and stakeholders by forcing research into a new mode of engagement with the broader world. It requires new incentives for individuals as well as institutions.

Acknowledgments

We are grateful for helpful comments and suggestions from other members of the CPWF management team: Alain Vidal, Tonya Schuetz, Michael Victor and Tim Williams; CPWF Basin Leaders: Olufunke Cofie, Alan Duncan, Kim Geheb, Pamela George, Simon Langan, Miguel Saravia; Andy Hall; IWMI colleagues: Jeremy Bird and Elizabeth Weight; and two anonymous reviewers.

Notes

1 Platform in the sense of, "A place or opportunity for public discussion" (Merriam-Webster, 1989).
2 The World Bank defines social capital as "the norms and networks that enable collective action. Increasing evidence shows that social cohesion—social capital—is critical for poverty alleviation and sustainable human and economic development." go.worldbank.org/X17RX35L00 (accessed 27 January 2014).
3 Agency is "the capacity, condition, or state of acting or exerting power" (Merriam-Webster, 1989). It is commonly applied in sociology to individuals or groups that act independently.
4 PNxx, MKx, ANx and Gangesx refer respectively to Phase 1, Mekong BDC, Andes BDC and Ganges BDC projects. Details are in the Appendix.

References

Alvarez, S., Douthwaite, B., Thiele, G., Mackay, R., Córdoba, D. and Tehelen, K. (2010) 'Participatory impact pathways analysis: A practical method for project planning and evaluation', *Development in Practice*, vol. 20, no. 8, pp. 946–958.
Candelo, C., Cantillo, L., Gonzalez, J., Roldan, A. and Johnson, N. (2008) 'Empowering communities to co-manage natural resources: Impacts of the *Conversatorio de Acción Ciudadana*', *Proceedings of the CGIAR Challenge Program on Water and Food 2nd International Forum on Water and Food, Addis Ababa, Ethiopia, 10–14 November 2008*, CGIAR Challenge Program on Water and Food, Colombo, Sri Lanka.
CGIAR (2000) *Integrated natural resource management research in the CGIAR*, a brief report on the INRM (integrated natural resource management) workshop held in Penang, Malaysia, 21–25 August 2000, Consultative Group on International Agricultural Research Secretariat.
CGIAR (2001) Mid-term meeting (2001) *Designing and managing change in the CGIAR*. Report to the Mid-Term Meeting, 21–25 May 2001 Durban, South Africa, world bank.org/html/cgiar/publications/mtm01/mtm0105.pdf (accessed 30 November 2013).
CGIAR (undated a) *History of CGIAR*, cgiar.org/who-we-are/history-of-cgiar/ (accessed 30 November 2013).
CGIAR (undated b) *CGIAR Research Programs*, cgiar.org/our-research/cgiar-research-programs (accessed 30 November 2013).
Clayton, T. (2013) 'Learning is engagement', Agriculture and Ecosystems Blog wle.cgiar.org/blogs/2013/07/11/learning-is-engagement/ (accessed 30 November 2013).
Collis, W., Sultana, P., Barman, B. and Thompson, P. (2011) 'Scaling out enhanced floodplain productivity by poor communities—aquaculture and fisheries in

Bangladesh and eastern India', *Proceedings of the CGIAR Challenge Program on Water and Food 3rd International Forum on Water and Food, Tshwane, South Africa, 14–17 November 2011*, CGIAR Challenge Program on Water and Food, Colombo, Sri Lanka.

CPWF Consortium (2002) *CGIAR Challenge Program on Water and Food Full Proposal*, r4d.dfid.gov.uk/Output/182299/Default.aspx (accessed 14 March 2013).

Cullen, B. (2013) *Communal grazing land management in Fogera: Lessons from innovation platform action research*, Nile Basin Development Challenge–Rainwater Management for Resilient Livelihoods, nilebdc.org/2013/01/18/ip-fogera-lessons/ (accessed 30 November 2013).

Douxchamps, S., Ayantunde, A. and Barron, J. (2012a) *Evolution of agricultural water management in rainfed crop-livestock systems of the Volta Basin*, CPWF Research for Development (R4D) Series 04, CGIAR Challenge Program on Water and Food, Colombo, Sri Lanka.

Douxchamps, S., Ayantunde, A. and Barron, J. (2012b) 'Evolution of agricultural water management in smallholder crop-livestock systems of the Volta Basin', *Workshop on Rainfed Production under Growing Rain Variability: Closing the Yield Gap*, 26–31 August 2012, Stockholm International Water Institute (SIWI), Stockholm, Sweden.

Duncan, A. (2012) *Project Nile 5: On coordination and multi-stakeholder platforms*, CPWF Six-Monthly Project Reports for April 2012–September 2012, unpublished.

Escobar, G. (2012) *Project Andes 1: Agua en los Andes, benefit-sharing mechanisms to improve water productivity and reduce water-related conflict in selected basins*, CPWF Six-Monthly Project Reports for April 2012–September 2012, unpublished.

Geheb, K. (2012) *Project Mekong 5: Coordination and change project*, CPWF Six-Monthly Project Reports for April 2012–September 2012, unpublished.

George, P. and Meisner, C. (2013) *CPWF BDC Progress Reports—Ganges*, November 2012–April 2013. Challenge Program on Water and Food, unpublished.

Johnson, N. (2009) *Sustaining inclusive collective action that links across economic and ecological scales in upper watersheds*, CPWF Project Report Series PN20, CGIAR Challenge Program on Water and Food, Colombo, Sri Lanka.

Kam, S. P. (2010) *Valuing the role of living aquatic resources to rural livelihoods in multiple-use, seasonally inundated wetlands in the Yellow River basin of China, for improved governance*, CPWF Project Report Series PN69, CGIAR Challenge Program on Water and Food, Colombo, Sri Lanka.

Kemp-Benedict, E., Cook, S., Allen, S. L., Vosti, S., Lemoalle, J., Giordano, M., Ward, J. and Kaczan, D. (2011) 'Connections between poverty, water and agriculture: evidence from 10 river basins', *Water International*, vol. 36, no. 1, pp. 125–140.

Lebel, L., Bastakoti, R. C. and Daniel, D. (2010) *Enhancing multi-scale Mekong water governance*, CPWF Project Report Series PN50, CGIAR Challenge Program on Water and Food, Colombo, Sri Lanka.

Mapezda, E., Amede, T., Geheb, K., Peden, D., Boelee, E., Demissie, T., van Hoeve, E. and van Koppen, B. (2008) 'Why gender matters: Reflections from the livestock-water productivity research project', *Proceedings of the CGIAR Challenge Program on Water and Food 2nd International Forum on Water and Food, Addis Ababa, Ethiopia, 10–14 November 2008*, CGIAR Challenge Program on Water and Food, Colombo, Sri Lanka.

Mbabu, A. and Hall, A. (eds) (2012) *Capacity building for agricultural research for development: Lessons from practice in Papua New Guinea*, UNU-MERIT, Maastricht.

Merriam-Webster (1989) *Webster's Ninth New Collegiate Dictionary*, Merriam-Webster, Springfield, MA.

Moore-Lappe, F., Collins, J., Rosset, P. and Esparza, L. (1986) *World Hunger: Twelve Myths*, Grove Press, New York NY.

Ormerod, P. (2005) *Why most things fail: Evolution, extinction and economics*, Wiley, Hoboken NJ.

Özgediz, S. (2012) *The CGIAR at 40: Institutional evolution of the World's premier agricultural research network*, CGIAR, Washington DC, library.cgiar.org/bitstream/handle/10947/2761/cgiar40yrs_book_final_sept2012.pdf (accessed 25 May 2014).

Pukinskis, I. and Geheb, K. (2012) *The Impacts of dams on the fisheries of the Mekong*, State of Knowledge Papers Series SOK 01, CGIAR Challenge Program on Water and Food, Vientiane, Lao PDR.

Quintero, M. (2012) *Project Andes 2: Assessing and anticipating the consequences of introducing benefit sharing mechanisms (BSMs)*, CPWF Six-Monthly Project Reports for April 2012–September 2012, unpublished.

Saravia, M. (2012) *Project Andes 4: Andes Coordination Project*, CPWF Six-Monthly Project Reports for April 2012–September 2012, unpublished.

Scoones, I. (1996) *Hazards and opportunities: Farming livelihoods in dryland Africa. Lessons from Zimbabwe*, Zed Books, London.

Sheriff, N., Joffre, O., Hong, M. C., Barman, B., Haque, A. B .M., Rahman, F., Zhu, J., Nguyen, H. van, Russell, A., Brakel, M. van, Valmonte-Santos, R., Werthmann, C. and Kodio, A. (2010) *Community-based fish culture in seasonal floodplains and irrigation systems*, CPWF Project Report Series PN35, CGIAR Challenge Program on Water and Food, Colombo, Sri Lanka.

TAC Secretariat (1993) *Progress Report by the Livestock Strategy Working Group, Mid-Term Meeting, May 23–28, 1993, San Juan, Puerto Rico.*

TAC (1993) *The ecoregional approach to research in the CGIAR. Report of the TAC/Center Directors Working Group*. TAC Secretariat and Food and Agriculture Organization of the United Nations, March 1993, ufdc.ufl.edu/AA00008181/00001/1x (accessed 30 November 2013).

Thurow, R. and Kilman, S. (2009) *Enough: Why the world's poorest starve in an age of plenty*, PublicAffairs, New York, NY.

Toufique, K. A. and Gregory, R. (2008) 'Common waters and private lands: Distributional impacts of floodplain aquaculture in Bangladesh', *Food Policy*, vol. 33, no. 6, pp. 587–594.

van Rooyen, A. and Homann-Kee Tui, S. (2009) 'Promoting goat markets and technology development in semi-arid Zimbabwe for food security and income growth', *Tropical and Subtropical Agroecosystems*, vol. 11, pp. 1–5.

Waisbord, S. R. (2000) *Watchdog journalism in South America: News, accountability and democracy*, Columbia University Press, New York, NY.

Waisbord, S. R. (2001) *Family tree of theories, methodologies and strategies in development communication*, prepared for The Rockefeller Foundation, communicationforsocial change.org/pdf/familytree.pdf (accessed 10 December 2013).

Woolley, J., Ribaut, J., Bouis, H. and Adekunle, A. (2009) *The CGIAR's Challenge Program experiences: A critical analysis*, Contribution to the first meeting of the Consortium Planning Team with the Alliance Executive and Deputy Executive (17–20 February 2009), cgspace.cgiar.org/handle/10568/35477 (accessed 29 April 2014).

Woolley, J. and Douthwaite, D. (2011) *Improving the resilience of agricultural systems through research partnership: A review of evidence from CPWF projects*, CPWF Impact Assessment Series 10, CGIAR Challenge Program on Water and Food, Colombo, Sri Lanka.

8 From research outputs to development outcomes— selected stories

Terry Clayton[a]* and *Michael Victor*[b]

[a]CGIAR Challenge Program on Water and Food—Mekong CPWF, Vientiane, Lao PDR; [b]CGIAR Research Program on Water, Land and Ecosystems WLE, Vientiane, Lao PDR; *Corresponding author, clayton@redplough.com.

Introduction

The CGIAR Challenge Program on Water and Food (CPWF) was a time-bound program with an end date of December 2013. The project adopted research-for-development (R4D) approaches only in its second Phase (2009–2013). This complicated its efforts to get from outputs to outcomes because decision-making is often a long process, vulnerable to multiple influences and driven by personalities. Credible and relevant research can increase the likelihood—but does not ensure—that outcomes will be achieved.

The overall purpose of the CPWF has been to address "wicked problems"[1] of water and food in complex adaptive systems. The project used a dynamic approach (R4D) to translate research outputs to development outcomes, defined as changes in policy or practice. To get from outputs to outcomes, the CPWF has used engagement strategies to influence decision-maker knowledge, attitudes and skills and facilitate negotiation.

Who are the decision-makers whose changes in policy or practice comprise outcomes? They can be officials making decisions on government policy. But they can also be community leaders deciding on collective management of resources, parties to negotiation about resource use, or development assistance agencies deciding on investment priorities. They can also be individual farm families deciding on agroecosystem management, or researchers deciding on research priorities.

Earlier chapters touched on examples of success in translating outputs to outcomes. This chapter presents some of these examples as outcome stories. The selection of stories is incomplete but covers five categories of decision-making: policy change, community resource management, negotiations regarding resource use, development investment and research priorities.

Introduction to the stories

It is a basic principle of R4D not to do things alone or in isolation of a larger context. At different times and places CPWF projects extended the work of others, filled gaps, raised questions, and sometimes challenged accepted wisdom. All the CPWF projects outlined in the stories that follow were a part of a bigger picture, which sometimes involved many people. As the stories illustrate, much of the CPWF's success was due to the efforts of its partners.

The stories are snapshots told from the CPWF perspective to illuminate some of the principles discussed in the preceding chapters. Getting to outcomes involved many people over long periods of time. In the following, we highlight particular strands of the stories. In so doing, much of the complexity has been simplified and much of the detail lost. For the benefit of interested readers who want to learn more about the bigger picture, the documents from which these snapshots are drawn are cited at the end of each story.

Research outputs and policy change

Addressing public health issues in urban vegetable farming in Ghana (PN38, PN51)[2]

Using diluted wastewater is a common practice in urban and peri-urban agriculture in sub-Saharan Africa and other low-income countries and is known to pose health risks. Vegetable growers in and around Accra felt their livelihoods were threatened by a city by-law banning the use of wastewater for irrigation. Two CPWF projects (PN38, PN51) were able to build on past work by local universities in an effort to improve producer livelihoods without risking consumer health.

Working with peri-urban farmers, researchers came up with simple but effective solutions such as low-cost water-treatment methods and safer irrigation practices. Post-harvest measures entailed washing methods and the use of disinfectants. Sedimentation ponds and sand filtration were tested and found to reduce helminth eggs to acceptable levels. Lowering the watering height and using a spray head on watering cans also reduced helminth egg and coliform counts.

The project helped establish strong working relationships with farmer organizations and networks of farmers and food sellers. It led to follow-up projects with WHO, International Development Research Centre, and United Nations Food and Agriculture Organization, and the founding of the Ghana Environmental Health Platform. The Platform continues the work started by local universities and CPWF partners. Project researchers also provided inputs to the WHO guidelines for wastewater use in agriculture.

In close collaboration with the Resource Centre on Urban Agriculture and Food Security, researchers initiated a revision of the Accra by-laws banning the use of wastewater. In 2010, the national irrigation policy was launched, which stated that, with certain precautions, the re-use of wastewater could be beneficial.

Citations relevant to this story are: Amoah et al., 2005; IWMI, 2005; Keraita et al., 2007; Abaidoo et al., 2009; IWMI, 2009; CPWF, 2012b.

Permits and pumping in West Bengal (PN42, PN60)[3]

For decades, the Indian Government had sought to intensify farming in the eastern Ganges by increasing the area of farmland irrigated with groundwater. But it was almost impossible for smallholders in West Bengal state in the eastern Ganges to access shallow groundwater for irrigation. The state government required farmers to obtain a permit, which it justified as a means to avoid over-exploiting groundwater. In practice, however, the system allowed rent-seeking by corrupt officials. Even with a permit, farmers still had to pay the cost of connecting to the electricity supply, which included cabling, transformer, poles and installation, plus the cost of the pump itself. The cost far exceeded the ability of smallholders to pay.

Researchers in project PN42 carried out hydrological studies that identified 301 "safe blocks" in West Bengal that had high annual recharge of the aquifer so that there was no risk of depleting it. Researchers then analyzed the permit system and documented how costly it was. Researchers in PN60 showed the state government that the permit system was an obstacle to the national government priority to increase productivity. The state government then changed policy to allow farmers in the 301 safe blocks to pump without permits as long as the pumps were less than 5 horsepower and discharged less than 30 m^3/hour. The state electricity authority rationalized connection fees to a fixed fee of 1000–30,000 rupees (US$16–US$475, September 2013) depending on the connected load.

Irrigated farming by smallholders in West Bengal using small pumps has increased rapidly. Aditi Mukherji, the lead researcher, was awarded the inaugural Norman Borlaug Award for Field Research and Application for this work.

CPWF researchers identified safe zones where groundwater was not at risk, quantified the cost of the permit policy and engaged with policymakers to convince them of the need to change the policy. The research continued in the International Water Management Institute with support from the Gates Foundation. CPWF was just one partner, but it made key contributions to the innovation process that provided the evidence to convince policymakers of the need to change.

Citations relevant to this story are: Mukherji, 2008; Mukherji et al., 2009; Mukherji, 2012a; Mukherji et al., 2012.

Paying for environmental services in an Andean watershed:
Encouraging outcomes from conservation agriculture (PN22)

Concerns were mounting over the health and biodiversity of the Lake Fuquene watershed near Bogotá, which provides environmental services such as tourism, urban water supplies and flood control. Reclamation of land for cattle-raising had reduced the lake area, and runoff from crop production and cattle manure was polluting the lake and causing eutrophication.

To address these problems, local partners promoted a transition from traditional practices to conservation agriculture. Even after wide promotion, however, adoption remained limited. Farmers blamed this on a lack of financial resources for initial investment and a lack of technical knowledge, and because many farmers were producing on rented land.

Adoption of conservation agriculture picked up when researchers implanted a scheme for payment for environmental services with a revolving fund managed by farmers' associations. The fund provided credit to make an initial investment in conservation agriculture. To get credit, a farmer had to present an approved land-use plan. Ninety-seven percent of the farmers receiving credit kept to the agreed plan and, thanks to a resulting increase of the average farm income from US$1850 to US$2180 per hectare, 100 percent of the first round of loans was recovered.

Many people in Andean watersheds do not own land, and therefore cannot benefit from agriculture or compensation for environmental services. In Colombia, conservation agriculture had positive impacts on soil characteristics by improving stream-flow regulation and reducing sediments, while increasing farm income. Low-interest loans proved to be an effective mechanism to promote practices that reduced sediment load and increased carbon sequestration. Long-term investment in payment for environmental services schemes is often affected by unfavorable macro-economic changes because public investment is invariably diverted to more immediate priorities. The Colombian experience demonstrates, however, that in the short-term, payment for environmental services can serve as an effective entry point for conservation agriculture.

Citations relevant to this story are: Rubiano et al., 2006; Estrada et al., 2009.

Research outputs and community resource management

Corralling: A solution to improving livestock productivity in
pasturelands affected by termites (PN37)

A CPWF research team looked at options to improve livestock water productivity in pastureland that had been corralled off in Nakosongala on the cattle corridor of Uganda. Ethiopian colleagues reported that corralling cattle every night over a two-week period helped recover desertified grassland. This simple solution was a breakthrough to a problem that had eluded ecologists and put livestock keepers under scrutiny for their role in accelerating land degradation.

For the practice to work, the villagers had to agree to corral their animals at night for two weeks to obtain sufficient quantities of cattle manure before moving elsewhere. Soon after, vegetative pasture grass cover was restored and surface water runoff and evaporation was reduced. Less silt and sediment of tanks and water reservoirs resulted in improved water quality.

In response to the development of these new technologies, local communities passed by-laws to protect vegetation and water quality. The impact of traditional livestock corralling on runoff and soil erosion levels varied with scale, cropping patterns, land use and tenure arrangements of the pasturelands.

Termites are generally viewed as a destructive insect, but soil scientists know that termites promote soil fertility and infiltration when in balance with nature. A technical solution to the termite problem opened up opportunities for a more systems-oriented approach to improving livelihoods while protecting the environment from desertification and water degradation.

A wider research study highlighted the need for improvements in legislative structure and institutional arrangements and understanding the importance of private land tenure. It was also important to know when promoting community-based natural resources management was appropriate, and what market opportunities there were. Providing better veterinary services helped to increase livestock and water productivity.

Citations relevant to this story are: Alemayehu et al., 2008; Mpairwe et al., 2008; Mugerwa et al., 2008; Peden, 2008; Haileslassie et al., 2009; Peden, 2008.

Changing slash and burn for slash and mulch in Central America's drought-prone hillsides (PN15)[4]

Slash and burn is the traditional management of many tropical systems. Smallholders cut down and burn forest growth, with ash from the burn providing the soil with nutrients, but leaving the soil with no cover and vulnerable to erosion. They then grow several food crops in succession until yields fall because of declining fertility, then they move on to the next block of regrowth forest. As long as the cycle is long enough, 20 or more years, for the trees to harvest the slow accumulation of nutrients from primary soil minerals, the system is sustainable. But, because of increasing population, the cycles have become shorter and shorter, with fewer crops possible in each. In most places, this short-cycle slash and burn is no longer sustainable, especially in drought-prone hillsides in the sub-humid tropics, with low, unreliable yields and increasing soil degradation.

Farmers in northeast Nicaragua and southeast Honduras recognized the problem and sought a solution. Working with the United Nations Food and

Agriculture Organization (FAO) and other partners in the 1990s, they developed the Quesungual slash and mulch agroforestry system (QSMAS), named for the village where it was developed. As a community, the farmers resolved to proscribe burning. They harvested useful timber in the secondary forest, slashed the rest and planted their crops directly into the dried, slashed vegetation. The benefits were such that after just one season, adopting farmers increased the area of the system on their farms. Other farmers in the region followed suit. Now 90 percent of the 120 farmers in the watershed have stopped slash and burn and over 60 percent use QSMAS.

Project PN15 reviewed the data collected in the FAO QSMAS project 1995–2005. Based on the analysis, PN15 then carried out research to: (1) measure and quantify the ecological and economic consequences of adopting QSMAS; (2) identify technical and socio-economic factors that control adoption of QSMAS; and (3) foster scaling out of QSMAS to other parts of Honduras, Nicaragua and Guatemala.

Project PN15 showed that QSMAS improved soil fertility, increased soil microflora, increased soil water-holding capacity and, hence, increased the tolerance of crops to drought. Crop yields and, hence, family incomes increased. There was less erosion and less risk of landslides during heavy rain, and moreover there was less pressure on primary forest. During the strong El Niño drought in 1997, crops in the QSMAS suffered less drought damage than crops in slash-and-burn systems. In the following year there was little erosion and no landslides during Hurricane Mitch, which dumped up to 1600 mm of rain in four days in the area.

QSMAS had a huge impact on the livelihoods of smallholder farmers (those who farm only 1 hectare). The most dramatic effect was the increased tolerance of crops to drought at the end of the rainy season, when rainfall is irregular and crops are filling their grain. Moreover, the restored forest environment protects the soil and buffers crop-production system and makes food supply resilient to unusually dry or wet years. Farmers now have a more sustainable source of wood supply for fuel and construction.

PN15 did not discover or invent QSMAS, but took its place in the longer-term trajectory of a self-propelled innovation process. Work on QSMAS continues in other forms and with other sources of funding.

Citations relevant to this story are: Pavon et al., 2006; Castro et al., 2008; Castro et al., 2009; CPWF, 2012a.

Opportunity in adversity: Collective fish culture in seasonal floodplains in Bangladesh (PN35)

Past efforts to increase the productivity of seasonal floodplains focused on increasing water productivity during the dry season when farmers were able to plant food crops. A community fishers' society in Beel Mail entered into a leasing arrangement with the local authorities which allowed them to fish during the flood season when the land is inundated. Before the society was set up, households fished individually and competed with one another for the catch.

Householders learned to work together to coordinate their activities and to protect the fish stock. Through a benefit-sharing arrangement, landowners and fishers both receive a share of the net profit. The society became financially more stable and strengthened its ability to raise funds for next year's purchase of fingerlings and the fencing needed to prevent fish escaping from the ponds. The lease was extended through 2012, with an option for a further three years.

Project PN35 sought to understand the socio-economic conditions that allowed the community-managed fish culture in Beel Mail and elsewhere to function successfully. It found that improved technology was a component, but practices that gave more equitable access to the fisheries, including the landless poor who were allowed unlimited rights to fish with hook and line, were important at the community level. Their success depended on communities enforcing their own rules, and maintaining lease rights to seasonally flooded areas used for fish production. Land tenure was a key component. Growing competition from private investors is likely to be an issue for the future.

Citations relevant to this story are: Sheriff et al., 2010; Collis et al., 2011; CPWF, 2012f; WorldFish, 2013.

Research outputs and negotiations about resource use

Enhancing multi-scale water governance in the Mekong Basin (PN50) (PN67)

There are a number of outcome stories emerging from the work of several projects working on water governance and multi-stakeholder platforms in the Mekong. The stories illustrate the slow but progressive build-up of information, understanding and capacity required in this action arena. One of the lessons from the work from the Mekong Program on Water, Environment and Resilience (M-POWER), applicable anywhere, is that public participation programs are not a panacea. Governments and other organizations often believe that participation is a cost not a benefit. Participation can also legitimize otherwise flawed processes and decisions while sidelining issues of gender and equality. The following stories illustrate that engagement backed by credible, salient and legitimate information can lead to changes in investment behavior, policy implementation and practices.

M-POWER is a network of collaborating scholars undertaking action-based research, facilitated dialogues and knowledge brokering to improve water governance in the Mekong region. It does this by supporting sustainable livelihoods, healthy communities and ecosystems. M-POWER was established in 2004 with the goal of democratizing water governance in the Mekong region through research and dialogue.

In Phase 1, the CPWF, together with the French Ministry of Foreign Affairs and IFAD funded a fellowship program partnered with M-POWER. The fellowship program was an important capacity-building initiative. It established a cohort of governance scholars within the Mekong region with 60 fellows. They will continue to have a constructive influence on water policy and decision-making in the region for decades to come.

The CPWF Project PN50 "Enhancing multi-scale Mekong water governance" was a flagship activity of M-POWER. The main goal was to help improve livelihood security and human and ecosystem health in the Mekong region through democratizing water governance. Fellows pursued this goal through critical research and direct engagement with stakeholders. The stakeholders were involved in managing fisheries, floods, irrigation, hydropower, watersheds, urban water works and integrated water management at various scales. In each policy domain, fellows identified common, shared, problems with current patterns of governance and made suggestions on how they could be addressed. Many of the activities supported coordination and collaboration among individuals working in six countries. The network grew substantially as a result of the M-POWER Fellowship program.

Policy dialogue

There was progress in strengthening local representation and the value of local inputs into planning and implementation recognized by central government agencies. Given the local styles of governance, this was no small accomplishment. The quality of deliberative processes improved. The body of event-convening and multi-stakeholder platforms grew and became an important resource on which other projects and practitioners could draw. A constructive interplay among institutions both horizontally and vertically gave important roles for engaged scholarship to link non-state and state actors at various levels.

Fisheries

It is now no longer possible to talk about dams and not talk about fisheries. There is an entrenched narrative of doom and crisis regarding the region's fisheries, however, that underpins policy, research and debate. Evidence of the potential adverse impacts of infrastructure on fisheries is now acknowledged. This initiative contributed to Xayaburi dam developers incorporating a fish pass in the controversy over constructing the dam.

Flood management

Flood management policies, measures and practices in the greater Mekong region are intended to reduce risks but often shift those risks on to vulnerable and disadvantaged groups. Government policymakers and dam developers now pay more attention to issues such as compensation for resettled villagers. The M-POWER flood working group undertook activities that helped establish flood and disaster management in the Mekong region as valid subjects for social, institutional and political analysis. On a more practical level, in An Giang Province in Vietnam's Mekong Delta, the People's Committee encouraged residents to explore and adapt to flood conditions. This led to several successful livelihood projects and brought "living with floods" policies closer to reality. These experiences were presented to the central government and used to revise the National Strategy on Disaster Prevention, Control and Mitigation in Vietnam to the year 2020.

Hydropower development

The M-POWER initiative contributed to national hydro-planning processes becoming more accessible to the public, both in terms of improved participatory processes, and in terms of improved accountability. The work addresses some of the shortcomings in the Hydropower Sustainability Assessment Protocol. The Second Mekong Forum on Water, Food and Energy in Hanoi in November 2012 showcased a number of M-POWER studies. The Forum was attended by senior government officials and private sector dam developers. The Third Forum in 2013 tabled new issues that emerged from this dialogue. The issues included managing the cumulative impact of dam cascades, sediment flows, increasing the role of private sector financing and corporate social responsibility. Gathering such a diverse group in one forum to address topics previously discussed behind closed doors was a milestone in the Mekong.

Heading in the right direction

How problems and solutions are framed has an important influence on which policies and projects are pursued. It is apparent that pathways to influence are diverse and do not depend solely on expert advice or a rational comparison of policy options. Although water bureaucracies use the language of participation and integration, they are not much used in practice.

A key strength of M-POWER was its rapid response capacity. As opportunities arose to influence policy, members of the network quickly let each other know and organized coordinated responses. The mixture and coverage of the network allowed for a flexible mix of individuals and actions. M-POWER research was compiled and published in three volumes, which were widely distributed. In Phase 2, several on-going multi-partner projects were secured and many partners committed to continue working together, which is encouraging.

Citations relevant to this story are: Molle et al., 2009; Dore et al., 2010; Lebel et al., 2010; Dore et al., 2012; Pukinskis and Geheb, 2012; Sajor, 2012.

Companion modeling for integrated renewable resource management:
A new collaborative approach to create common values for sustainable
development (PN25)

Companion modeling (ComMod) was a collaborative modeling and simulation approach. It used an interactive process and mediating tools to support dialogue, shared learning, negotiation and collective decision-making. In addition to generating possible solutions to problems, the process helped strengthen the capacity of communities for adaptive management. The principle of the ComMod approach was to co-construct simulation tools that integrated different points of view and use them to examine problems of resource management.

In the Mekong Basin and the Himalayan highlands, users compete for water to serve various needs with different strategies, water-use practices and perceptions of the problem. Companion modeling was a way to incorporate many points of view into a collective decision-making process. Ideally, solutions needed to reconcile ecological and social dynamics. They also needed to improve communication, collective learning, coordination mechanisms and stakeholders' capacity for adaptive management and collective action. The question was, would ComMod work in different settings in Bhutan, Thailand and Vietnam?

The project led to improved communication and trust among multiple stakeholders, for example between forest authorities and villagers in Thailand and among different groups of herders in Bhutan, and between rice and shrimp farmers in Vietnam.

Project activities led to new regulations on the use of irrigation water in the Salaep catchment in Thailand. In Bhutan, a resource management committee was established in the Lingmuteychu watershed where the Kengkhar villagers agreed to coordinate the use of water tanks. In coastal Vietnam, downstream shrimp farmers and upstream rice growers compromised on the timing of the intake of saline water at an important sluice.

Citations relevant to this story are: Ruankaew et al., 2010; CIRAD, 2012; CPWF, 2012e.

Research outputs and development investment

Goats and fodder in Zimbabwe (Limpopo 3)[5]

In Zimbabwe, there was an unmet demand for goat meat in local and national markets, yet this demand was not reflected in prices at the farm gate. Most goats were produced by poor households in marginal areas where they played an important role in household livelihoods. Most goat carcasses were poor quality and as many as 20 percent or more goats starved during the dry season because of lack of fodder. As a result, many goats were often sold in distress sales where the producer had to accept whatever low price the dealer offered.

The solution was a system of formal goat auctions on defined dates promoted by an innovation platform of farmers, dealers, researchers and other stakeholders. The auctions were held in sale yards, which provided scales and pens for small animals. Producers saw prices increase from US$10 to US$50–60 a head as buyers now had to bid for each lot and pay more for quality. The change induced producers to improve management to improve carcass quality, with innovations such as fencing, animal health and improved feed. With improved management, mortality fell and is now about 10 percent, and there are many fewer distress sales.

The demand for improved feed is ironic. For many years researchers had promoted pastures as feed for goats in the dry season, but farmers had little incentive to take up the technology. With farm income from goats increasing from US$10 to about US$200 per year, farmers now have resources and incentive to improve the feed for their animals in the dry season and seek research support to implement pasture improvement.

The innovation platform to create the auction system was funded by Germany, the EU and the Swiss Agency for Development and Cooperation. The CPWF project Limpopo 3 worked with researchers to identify improved dry-season fodders that the farmers now sought. The CPWF provided support that hastened progress of an innovation platform that was already in place.

Citations relevant to this story are: van Rooyen and Homann-Kee Tui, 2008; van Rooyen and Homann-Kee Tui, 2009; van Rooyen, 2012.

Safeguarding livelihoods in the GaMampa wetlands in the Limpopo River basin (PN30)

Long-term sustainable management and conservation requires that farmers participate as co-managers of their resources. Working with people in the community, researchers developed and applied a trade-off-based framework for making decisions about allocations of wetland resources for specific uses, including agriculture. They used WETSYS, a trade-off model, for assessing the costs and benefits of different uses of wetlands in a modeling exercise in the GaMampa wetland, South Africa (Morardet and Masiyandima, 2012). The exercise helped people in the local community and other stakeholders better understand the trade-offs involved in clearing reeds for preparing new areas for cropping. The model estimated that the GaMampa wetland could contribute US$210 per household per year to the livelihoods of the local communities, a sevenfold increase in their current income.

Public officials were still involved with the community more than a year after fieldwork ended. Engaging government officials responsible for natural resource management helped ensure that local concerns are incorporated into program management decisions. With support of the Landcare Unit of the Limpopo Department of Agriculture, the community obtained financial support from the United Nations Development Program to help them continue to manage their wetland resources.

Citations relevant to this story are: Morardet et al., 2010; CPWF, 2012h.

Enhancing water productivity and improving livelihoods through drip irrigation and better market integration in Cambodia (SG 502)

In Cambodia, the CPWF used a small grants program to design a strategy to enable smallholders to benefit from effective market participation. This was an International Development Enterprises (IDE) project in partnership with the World Vegetable Center and the International Fertilizer Development Center. The project trained farm business advisors who encouraged farmers to participate, guided them in their cropping decisions and connected them to suppliers of irrigation drip-kits and fertilizer briquettes. They also helped farmers to establish market links for their products. Farmer incomes more than doubled as a result.

The market for vegetables is sustained by the increasing demand in urban markets for high-value crops, so more producers do not cause a supply surplus, which would result in decreased prices. Before the project, vegetable production averaged only 9 percent of family income. After only one year of the project, vegetables accounted for 19 percent. Farmers who combined drip irrigation with deep placement of fertilizer to grow high-value crops increased their average net income by 33 percent.

The original assumption was that poor water management was the major constraint in commercial farming in the project area, and drip irrigation was viewed as a solution. Researchers found, however, that under traditional farming practices, it was not cost effective on its own and that the benefits of improved water management could be enhanced with attention to soil fertility, crop selection and technical advice to farmers.

The research team recruited respected farmers who were trained as farm business advisors and served as mobile retailers of horticultural products and services, and providers of technical advice. After the project ended, the farm business advisors continued to supply inputs and give technical advice to farmers, even after project stipends had been phased out. The continuing services were funded by the margins that the advisors charged on the products they sold.

Citations relevant to this story are: Palada et al., 2008; CPWF, 2012c.

The multiple-use water services (MUS) project (PN28)

Many water infrastructure projects are designed around a single use only, for example, irrigation or domestic use. However, rural families usually prefer to use water systems for multiple uses, for example irrigation and domestic use and livestock watering. Rural livelihoods can be improved by designing water systems from the beginning to accommodate multiple uses of water from multiple sources.

Multiple use is not a new idea but requires a major shift in thinking among practitioners in what are now segregated water sectors (e.g., sanitation, drinking water, hydropower, irrigation, etc.). In addition to more time and cost, MUS put engineers and managers into a complex new world they know little about and for which they are not trained. Practitioners have to engage with local users to determine current and future needs and then design, build and manage the necessary water infrastructure. MUS need new professional training curricula and new policies, regulations and institutions.

The project implemented MUS and scaled them up at intermediate, national and global levels in eight countries with 150 partners. In the eight countries, action research highlighted the benefits people derive from MUS designed for a single purpose. Research showed that developing multiple water sources and increasing water availability from 20 liters (the minimum recommended by WHO) to 100 liters per capita per day, raised yearly income derived from water from US$40 to US$300 per household.

The core team set up learning alliances with partners in the basins of the Andes (Bolivia and Colombia), Indus-Ganges (India and Nepal), Limpopo (South Africa and Zimbabwe), Mekong (Thailand) and Nile (Ethiopia), which enabled important local innovations. Global advocacy put MUS on the policy agendas of professional networks such as the World Water Forum, governmental and non-governmental water agencies, and rural development and financing organizations.

Nepal: Important impacts

The USA-based international NGO IDE has a long history in Nepal working with Winrock, local NGOs and governments at both national and local levels. IDE and its partners implemented learning alliances at national, district and local levels in over 80 projects, which is remarkable given the country's political and civil turmoil. The MUS project partnership complemented IDE-Winrock and their partners, which enhanced the credibility of MUS and learning alliances in Nepali eyes.

Learning alliances in Nepal helped break down institutional barriers between the agricultural and domestic water supply sectors. The IDE-Winrock Project shifted from working with individual households to working with communities. IDE credits the MUS Project Leader with pushing them to take a more participatory and gender-balanced approach than they might otherwise have used.

The Nepal Department of Irrigation (NDI) had previously initiated a non-traditional irrigation project to reach more people. Introducing MUS provided quick and visible successes, which led to positive recognition within NDI. NDI has now adopted a practical approach to including domestic water services in its irrigation schemes, which IDE anticipates will soon become policy. Much of the support for MUS for domestic water came from demands from communities using local budgets. MUS schemes were approved by the government for investment on a cost-share basis.

South Africa, international acknowledgement leads to local acceptance

A learning alliance helped coordinate a number of action research projects with AWARD, a local NGO. Whilst the results showed the practical application of MUS, water sector managers in South Africa rarely read research papers or policy briefs. Successful advocacy relies on personal contact rather than documentation. A local champion within the alliance created some local awareness, but not sufficient for wider adoption.

At the Fourth World Water Forum in Mexico in 2006, several key South Africans were involved in a special MUS workshop, the results of which were presented to the Department of Water Affairs and Forestry. International acceptance of South African research provided strong grounds for acceptance of MUS, and the African Development Bank provided funding. Researchers who continued to work on these issues were optimistic that MUS will be incorporated into legislation as a result of on-going debate about merging the Water Services Act and the National Water Act.

Research outputs and research priorities

Water control and farm intensification in the coastal Ganges (Ganges 1–5)

Water control in the lower Ganges in coastal Bangladesh is a critical issue. During the wet season the land was flooded by the river. The land also flooded from the sea during cyclones. During the dry season the river flow was negligible and seawater intruded back up the river. To protect the land from flooding, areas called polders[6] of tens to hundreds of hectares were surrounded by earthen walls 1–2 m high. The walls, called dykes, protected the polder from flooding in the wet season and from saltwater in the dry season. But farmers could only grow one crop of rice during the monsoon wet season because there was no freshwater to irrigate the polders in the dry season when they were surrounded by saline water.

Projects PN10 and Ganges 2 evolved from "a lack of fresh water to intensify cropping" to become "a lack of integrated water management within polders to enable intensified cropping." This, however, overlooked the need for infrastructure to manage (store) water from the wet-season floods for irrigation in the dry season. Lands inside polders are uneven, with low spots and high spots, complicating attempts at drainage and irrigation. The project therefore evolved to create water-control infrastructure to manage water within the polders. This allowed land cropped in the wet season—regardless whether in low spots or high spots—to be drained after harvest and to be sown with dry-season crops and irrigated with stored freshwater.

The projects identified a number of possibilities depending on the local river flow, which controls the salinity of the river water. Where it is possible to store water from the wet-season flood, farmers could grow two consecutive crops of rice, or wet-season rice followed by a high-value dry-season crop. This system depended on storing freshwater and then closing the intake to the polder before the river became too saline. Where storage was not possible, rice in the wet season could be followed by shrimp or fish in the dry season using brackish water from the river, admitted early in the dry season before the river becomes too salty. In some places, there was no problem because the river water is fresh almost year round, which allowed triple cropping with rice, or rice rotated with other crops.

The new contribution of Project Ganges 3 was the evolution of its objectives as the nature of the problems became clearer. What was needed were new ways to manage crops and more importantly, new institutions to manage water. The government had to rationalize its strategies of investment in polders and their maintenance, and the design of infrastructure within the polders. Also critical was the management of water entry at the sluice gates. As researchers understood the problem better, they changed the design and priorities of the research to meet the development challenge.

Improving resilience among fishers in the Niger Basin (PN72)

In Africa, small-scale inland fisheries are important to the livelihoods of the poor, contributing both income and food security for millions of households. These inland fisheries are difficult to assess and manage because of complex multi-species, multi-gear fishing and large numbers of fishers operating within the informal sector.

In the inland delta of the Niger, researchers worked with communities to assess their livelihood strategies and develop adaptive management solutions to address problems. The research required new ways of looking at the problems, and researchers used the concept of resilience to understand vulnerability and how resources could be better managed.

Guided by the research team, fishers identified resilience indicators and interventions. The decision-making process was enhanced by the creation of community-level committees. The committees' composition ensured gender equity and reduced the likelihood of control by the most powerful individuals and households in the communities.

The project challenged the conventional view that research should focus on the natural resource base. Although the communities acknowledged that depleted and fluctuating fish stocks affected their livelihoods, they identified more fundamental causes of vulnerability. These included food insecurity, exposure to water-borne diseases and lack of access to cash and micro-credit facilities.

An important lesson was that the most productive interventions for promoting sustainable resource use may lie outside the natural system. Hence, interventions improved access to drinking water by rehabilitating boreholes, renovated flood control infrastructure, improved school facilities and created micro-credit facilities. Strengthening rural household resilience was as important a goal as adopting specific technologies.

Citations relevant to this story are: Béné et al., 2009; Ogilvie et al., 2010; CPWF, 2012d.

A final story that crosses categories (PN22 and Andes 1–4)

The final story, from the Andes, describes an example of a group of projects with outcomes that bridged different categories. The categories were policy change, community resource management, negotiations, development investment and research priorities.

Basins in the Andes that flow from east to west are short and steep. In Ecuador and Peru, there is abundant rainfall upstream but little rainfall downstream. Water use is concentrated downstream, while poverty is concentrated upstream. There are several upstream–downstream links that affect different groups with different interests in different ways (see Box 5.5 in Chapter 5). The principal link is that land and water management upstream affects water quality and other ecosystem services (ESSs) for downstream users. Everyone could be better off if institutional mechanisms were established to share water-related benefits and costs.

Benefit sharing means ESS beneficiaries paying ESS providers for the ESSs they provide. Cost sharing means making payments to provide ESSs by promoting better land and water management upstream. Benefit-sharing mechanisms (BSMs) are the means for downstream water users to make the payments. They encourage upstream management that gives positive externalities (and discourage practices with negative externalities). BSMs can provide incentives to use improved technical practices such as replacing intensive hillside tillage with no-till agriculture, introducing tree crops, and many others.

Technical innovation required institutional change, for example to create trust funds for collecting and making payments. The policy context was also favorable because many Andean countries give high priority to maintaining alpine ecosystems and to reducing poverty in highland communities.

Research outputs were important in taking decisions on the following questions:

- How should payments to ESS providers be targeted? (Hydrological modeling to identify upstream areas providing ESSs.)
- How should payments be used? (Research on upstream ecosystem conservation measures and social development projects.)
- How much should each class of ESS beneficiary contribute? (Estimate the economic value of watershed services for different downstream ESS users—used as reference values in negotiations about contributions to an ESS investment fund.)

There were legal and institutional bottlenecks that hindered implementation of institutional innovations for BSMs. Researchers participated in developing legislation to address some of these bottlenecks:

- Issues in transferring voluntary contributions from urban water user to an independent BSM investment fund;
- Issues in transferring public resources of local governments to these funds;
- How to make voluntary contributions legally binding;
- Financial independence and transparency of BSM investment funds;
- Lack of guidelines on how to establish new institutions for operating BSM investment funds; and
- Lack of institutional structure for integrated watershed management.

Modeling was important to identify which hydrological units in watersheds had the highest potential for change. It provided poverty profiles of farmers and identified the natural resources they controlled and for which they could receive payments. It also quantified the socio-economic benefits that land use changes could bring.

The team used the soil and water assessment tool (SWAT) and ECOSAUT (Box 8.2), a model for social, economic and environmental evaluation of land use. The team collected the necessary data from a wide range of sources. To determine changes in competitiveness, the team used an approach based on the policy analysis matrix (PAM). PAM allows the use of secondary data and requires the calculation of social values (shadow prices) for inputs and outputs. An optimization model such as ECOSAUT can be used to determine the shadow price of the land and labor under different technological alternatives.

The CPWF, together with CIAT, CONDESAN and other partners, played key research and engagement roles since 2005 in the development of BSMs. CONDESAN's work in coordinating research on BSMs in the Andes Basin Development Challenge was especially important. It was founded in 1993 as a spin-off of the CGIAR Ecoregional Program in the Andes. It became an important platform for natural resource management and sustainable development, particularly in water and watershed management. Its members include the Andean Community of Nations, the Spanish Agency for International Cooperation for Development and the Swiss Agency for Development and

---

Box 8.1 The models

The team combined information about topography, soils, weather and land use for simulation using the ArcView-SWAT interface. They incorporated data on soil characteristics, daily climate data, and delineated the watersheds using a digital elevation model. The project developed a simulation model, ECOSAUT, which uses linear programming to maximize farmers' net incomes under social, economical and environmental constraints. The model considers multiple land use systems and can quantify marginal benefits among a baseline and alternative land use systems.

Cooperation. CONDESAN is widely respected among governments, NGOs and community-based organizations. Its work will continue after CPWF funding has ceased.

We emphasize one particular example. Partnered with the Ministry of Environment of Peru, CPWF projects designed a BSM for the Cañete River basin, which the Ministry uses as a pilot project. The Cañete BSM established a trust fund to finance its activities (Quintero, 2012). CIAT and the CPWF partner, CONDESAN, worked with the Ministry and with public and private companies and NGOs. The objective was a BSM to provide equitable benefits from the use of ecosystems and benefit rural communities.

The findings are being scaled-out to over 30 river basins in Peru through a *Remuneration Mechanism for Ecosystem Services* hosted by the Ministry with CONDESAN's support. CIAT and the CPWF helped draft national legislation to evaluate ecosystem services. The legislation requires that valuation of water uses be included in all environmental impact studies for new public and private development projects.

Citations relevant to this story are: Estrada et al., 2009; Johnson et al., 2009; Quintero et al., 2009; Escobar and Estrada, 2011; CPWF, 2012g.

Notes

1 Wicked problems are those that are difficult to solve because their requirements are contradictory, changing, hard to reconcile and often not well understood (Rittel and Webber, 1973).
2 PN is the CPWF project number. A list of projects is in the Appendix.
3 This story was also featured in Chapter 5.
4 This story was also featured in Chapter 5.
5 This story was also featured in Chapter 5.
6 Polder: a piece of low-lying land reclaimed from the sea or a river and protected by dykes.

References

Abaidoo, R. C., Keraita, B., Amoah, P., Drechsel, P., Bakang, J., Kranjac-Berisavljevic, G., Konradsen, F., Agyekum, W. and Klutse, A. (2009) *Safeguarding public health concerns, livelihoods and productivity in wastewater irrigated urban and periurban vegetable farming*, CPWF Project Report Series PN38, CGIAR Challenge Program on Water and Food, Colombo, Sri Lanka.
Alemayehu, M., Peden, D., Taddesse, G., Haileselassie, A. and Ayalneh, A. (2008) 'Livestock water productivity in relation to natural resource management in mixed crop-livestock production systems of the Blue Nile River basin, Ethiopia', *Proceedings of the CGIAR Challenge Program on Water and Food 2nd International Forum on Water and Food, Addis Ababa, Ethiopia, 10–14 November 2008*, CGIAR Challenge Program on Water and Food, Colombo, Sri Lanka.
Amoah, P., Drechsel, P. and Abaidoo, R. C. (2005) 'Irrigated urban vegetable production in Ghana: Sources of pathogen contamination and health risk elimination', *Irrigation and Drainage* vol. 54, no. S1, pp. S49–S61.

Béné, C., Kodio, A., Lemoalle, J., Mills, D., Morand, P., Ovie, S., Sinaba, F. and Tafida, A. (2009) *Participatory diagnosis and adaptive management of small-scale fisheries in the Niger River basin,* CPWF Project Report Series PN72, CGIAR Challenge Program on Water and Food, Colombo, Sri Lanka.

Castro, A., Rivera, M., Ferreira, O., Pavon, J., García, E., Amezquita, E., Ayarza, M., Barrios, E., Rondon, M. and Pauli, N. (2008) 'Quesungual slash and mulch agroforestry system improves rain water productivity in hillside agroecosystems of the sub-humid tropics', in E. Humphreys and R. S. Bayot (eds) *Proceedings of the International Workshop on Rainfed Cropping Systems, Tamale, Ghana,* CGIAR Challenge Program on Water and Food, Colombo, Sri Lanka.

Castro, A., Rivera, M., Ferreira, O., Pavón, J., García, E., Amézquita, E., Ayarza, M., Barrios, E., Rondón, M., Pauli, N., Baltodano, M. E., Mendoza, B., Wélchez, L. A. and Rao, I. M. (2009) *Quesungual slash and mulch agroforestry system (QSMAS): Improving crop water productivity, food security and resource quality in the subhumid tropics,* CPWF Project Report Series PN15, CGIAR Challenge Program on Water and Food, Colombo, Sri Lanka.

CIRAD (2012) *Annual Report: Companion modeling at sub-basin scale in the Mekong,* unpublished.

Collis, W., Sultana, P., Barman, B. and Thompson, P. (2011) 'Scaling out enhanced floodplain productivity by poor communities—aquaculture and fisheries in Bangladesh and eastern India', *Proceedings of the CGIAR Challenge Program on Water and Food 3rd International Forum on Water and Food, Tshwane, South Africa, 14–17 November 2011,* CGIAR Challenge Program on Water and Food, Colombo, Sri Lanka.

CPWF (2012a) *Abandoning slash and burn for slash and mulch in Central America's drought-prone hillsides,* Outcome Stories Series, CGIAR Challenge Program on Water and Food, Colombo, Sri Lanka.

CPWF (2012b) *Addressing public health issues in urban vegetable farming in Ghana,* Outcome Stories Series, CGIAR Challenge Program on Water and Food, Colombo, Sri Lanka.

CPWF (2012c) *Enhancing water productivity and improving livelihoods through drip irrigation and better market integration in Cambodia,* Outcome Stories Series, CGIAR Challenge Program on Water and Food, Colombo, Sri Lanka.

CPWF (2012d) *Improving resilience among small-scale fishers in the Niger River basin,* Outcome Stories Series, CGIAR Challenge Program on Water and Food, Colombo, Sri Lanka.

CPWF (2012e) *Learning to share water in the highlands of Bhutan using the companion modeling approach,* Outcome Stories Series, CGIAR Challenge Program on Water and Food, Colombo, Sri Lanka.

CPWF (2012f) *Opportunity in adversity: Collective fish culture in the seasonal floodplains of Bangladesh,* Outcome Stories Series, CGIAR Challenge Program on Water and Food, Colombo, Sri Lanka.

CPWF (2012g) *Paying for environmental services in an Andean watershed: Encouraging outcomes from conservation agriculture,* Outcome Stories Series, CGIAR Challenge Program on Water and Food, Colombo, Sri Lanka.

CPWF (2012h) *Safeguarding livelihoods in the GaMampa wetlands in the Limpopo River basin,* Outcome Stories Series, CGIAR Challenge Program on Water and Food, Colombo, Sri Lanka.

Dore, J., Molle, F., Lebel, L., Foran, T. and Lazarus, K. (2010) *Improving Mekong water resources investment and allocation choices,* CPWF Project Report Series PN67, CGIAR Challenge Program on Water and Food, Colombo, Sri Lanka.

Dore, J., Lebel, L. and Molle, F. (2012) 'A framework for analysing transboundary water governance complexes, illustrated in the Mekong Region', *Journal of Hydrology* vols 466–467, pp. 23–36.

Escobar, G. and Estrada, R. B. (2011) *Diversity of water-based benefits in the High Andes Range,* AN1 Working Paper 1, RIMISP.

Estrada, R. D., Quintero, M., Moreno, A. and Ranvborg, H. M. (2009) *Payment for environmental services as a mechanism for promoting rural development in the upper watersheds of the tropics,* CPWF Project Report Series PN22, CGIAR Challenge Program on Water and Food, Colombo, Sri Lanka.

Haileslassie, A., Peden, D., Gebreselassie, S., Amede, T., Wagnew, A. and Taddesse, G. (2009) 'Livestock water productivity in the Blue Nile Basin: Assessment of farm scale heterogeneity', *The Rangeland Journal* vol. 31, no. 2, pp. 213–222.

IWMI (2005) *The impact of wastewater irrigation on human health and food safety among urban communities in the Volta Basin—opportunities and risks,* CPWF Project 51 Progress Report for 2005, unpublished.

IWMI (2009) *PN51 Completion Report–Waste water irrigation—opportunities and risks,* unpublished.

Johnson, N., Garcia, J., Rubiano, J. E., Quintero, M., Estrada, R. D., Mwangi, E., Morena, A., Peralta, A. and Granados, S. (2009) 'Water and poverty in two Colombian watersheds', *Water Alternatives* vol. 2, no. 1, pp. 34–52.

Keraita, B., Konradsen, F., Drechsel, P. and Abaidoo, R. C. (2007) 'Effect of low-cost irrigation methods on microbial contamination of lettuce irrigated with untreated wastewater', *Tropical Medicine & International Health* vol. 12, no. s2, pp. 15–22.

Lebel, L., Bastakoti, R. C. and Daniel, R. (2010) *Enhancing multi-scale Mekong water governance,* CPWF Project Report Series PN50, CGIAR Challenge Program on Water and Food, Colombo, Sri Lanka.

Molle, F., Foran, T. and Kakonen, M. (2009) *Contested waterscapes in the Mekong Region: Hydropower, livelihoods and governance,* London, Earthscan.

Morardet, S., Masiyandima, M., Jogo, M. and Juizo, D. (2010) 'Trade-offs between livelihoods and wetland ecosystem services: An integrated dynamic model of Ga-Mampa wetland, South Africa', *Proceedings of LANDMOD2010,* at Montpellier, February 3–5, 2010.

Morardet, S. and Masiyandima M. (2012) *WETSYS, a system dynamic model to assess trade-off between wetland ecosystem services at local level.* WETwin project report. cemadoc.irstea.fr/exl-php/docs/PUB_DOC/30686/2012/mo2012-pub00037190_PDF.txt (accessed 17 April 2014).

Mpairwe, D., Mutetikka, G., Kiwuwa, S., Owoyesigire, B., Zziwa, E. and Peden, D. (2008) 'Options to improve livestock-water productivity (LWP) in the cattle corridor within the White Nile sub-basin in Uganda', *Proceedings of the CGIAR Challenge Program on Water and Food 2nd International Forum on Water and Food, Addis Ababa, Ethiopia, 10–14 November 2008,* CGIAR Challenge Program on Water and Food, Colombo, Sri Lanka.

Mugerwa, S., Mpairwe, D., Sabiiti, E., Muthuwatta, L., Kiwuwa, G., Zziwa, E. and Peden, D. (2008) 'Cattle manure and reseeding effects on pasture productivity', *Proceedings of the CGIAR Challenge Program on Water and Food 2nd International Forum on Water and Food, Addis Ababa, Ethiopia, 10–14 November 2008,* CGIAR Challenge Program on Water and Food, Colombo, Sri Lanka.

Mukherji, A. (2008) 'Spatio-temporal analysis of markets for groundwater irrigation services in India: 1976–1977 to 1997–1998', *Hydrogeology Journal* vol. 16, no. 6, pp. 1077–1087.

Mukherji, A., Villholth, K. B., Sharma, B. and Wang, J. (2009) *Groundwater governance in the Indo-Gangetic and Yellow River basins: Realities and challenges*, International association of Hydrologists Selected Papers, vol. 15, CRC Press, Oxford.

Mukherji, A. (2012a) 'Innovations in managing the agriculture, energy and groundwater nexus: Evidence from two states in India', *Workshop on Towards a Green Economy: The Water–Food–Energy Nexus,* 26–31 August 2012, Stockholm, Sweden, Stockholm International Water Institute (SIWI).

Mukherji, A. (2012b) *Project Ganges 3: Water governance and community-based management,* CPWF Six-Monthly Project Reports for April 2012–September 2012, unpublished.

Mukherji, A., Shah, T. and Bannarjee, P. S. (2012) 'Kick-starting a second Green Revolution in Bengal', *Economic & Political Weekly* vol. XLVII, no. 18, pp. 27–30.

Ogilvie, A., Mahe, G., Ward, J., Serpantie, G., Lemoalle, J., Morand, P., Barbier, B., Diop, A. T., Caron, A. and Namarra, R. (2010) 'Water, agriculture and poverty in the Niger River basin', *Water International* vol. 35, no. 5, pp. 594–622.

Palada, M., Bhattarai, S., Roberts, M., Baxter, N., Bhattarai, M., Kimsan, R., San, K. and Wu, D. (2008) 'Increasing on-farm water productivity through affordable microirrigation vegetable-based technology in Cambodia', *Proceedings of the CGIAR Challenge Program on Water and Food 2nd International Forum on Water and Food, Addis Ababa, Ethiopia, 10–14 November 2008,* CGIAR Challenge Program on Water and Food, Colombo, Sri Lanka.

Pavon, J., Amezquita, E., Menocal, O., Ayarza, M. and Rao, I. M. (2006) 'Application of QSMAS principles to drought-prone areas of Nicaragua: Characterization of soil chemical and physical properties under traditional and QSMAS validation plots in La Danta watershed in Somotillo', *CIAT Annual Report 2006: Integrated soil fertility management in the tropics,* CIAT-TSBF, Cali, Colombia.

Peden, D. (2008) 'Could 150 million thirsty livestock be efficient water harvesters? Nile Basin studies show how', *Collective Action News, Issue No. 4,* November 2008, Alliance of the CGIAR Centers.

Peden, D., Alemayehu, M., Amede, T., Awulachew, S. B., Faki, H., Haileslassie, A., Herrero, M., Mapezda, E., Mpairwe, D., Musa, M. T., Taddesse, G. and Breugel, P. van. (2009) *Nile Basin livestock water productivity,* CPWF Project Report Series PN37, CGIAR Challenge Program on Water and Food, Colombo, Sri Lanka.

Pukinskis, I. and Geheb, K. (2012) *The impacts of dams on the fisheries of the Mekong,* State of Knowledge Papers Series SOK 01, CGIAR Challenge Program on Water and Food, Vientiane, Lao PDR.

Quintero, M., Wunder, S. and Estrada, R. D. (2009) 'For services rendered? Modeling hydrology and livelihoods in Andean payments for environmental services schemes', *Forest Ecology and Management* vol. 258, no. 9, pp. 1871–1880.

Quintero, M. (2012) *Project Andes 2: Assessing and anticipating the consequences of introducing benefit sharing mechanisms (BSMs),* CPWF Six-Monthly Project Reports for April 2012–September 2012, unpublished.

Rittel, H. W. J. and Webber, M. M. (1973) 'Dilemmas in a general theory of planning', *Policy Sciences* vol. 4, pp. 155–169.

Ruankaew, N., Le Page, C., Dumrongrojwatthana, P., Barnaud, C., Gajaseni, N., Paassen, A. van and Trebuil, G. (2010) 'Companion modeling for integrated renewable resource management: A new collaborative approach to create common values for sustainable development', *International Journal of Sustainable Development and World Ecology* vol. 17, no. 1, pp. 15–23.

Rubiano, J., Quintero, M., Estrada, R. D. and Moreno, A. (2006) 'Multiscale analysis for promoting integrated watershed management', *Water International* vol. 31, no. 3, pp. 398–411.

Sajor, E. (2012) *Project Mekong 4: Water governance*, CPWF Six-Monthly Project Reports for April 2012–September 2012, unpublished.

Sheriff, N., Joffre, O., Hong, M. C., Barman, B., Haque, A. B. M., Rahman, F., Zhu, J., Nguyen, H. van, Russell, A., Brakel, M. van, Valmonte-Santos, R., Werthmann, C. and Kodio, A. (2010) *Community-based fish culture in seasonal floodplains and irrigation systems*, CPWF Project Report Series PN35, CGIAR Challenge Program on Water and Food, Colombo, Sri Lanka.

van Rooyen, A. and Homann-Kee Tui, S. (2008) 'Enhancing incomes and livelihoods through improved farmers' practices on goat production and marketing', *Proceedings of a Workshop Organized by the Goat Forum*, 2–3 October 2007, Bulawayo, Zimbabwe.

van Rooyen, A. and Homann-Kee Tui, S. (2009) 'Promoting goat markets and technology development in semi-arid Zimbabwe for food security and income growth', *Tropical and Subtropical Agroecosystems* vol. 11, pp. 1–5.

van Rooyen, A. (2012) *Project Limpopo 3: Farm systems and risk management*, CPWF Six-Monthly Project Reports for April 2012–September 2012, unpublished.

WorldFish (2013) *PN35: Research into use—Floodplain fisheries and aquaculture in Bangladesh and India*, Six-Monthly Progress Report (2nd Report), unpublished.

9 Messages and meaning

Larry W. Harrington[a]* and *Alain Vidal*[b]

[a]CGIAR Challenge Program on Water and Food CPWF, Ithaca, NY, USA;
[b]CGIAR Challenge Program on Water and Food CPWF, Montpellier,
France; *Corresponding author, lwharrington@gmail.com.

The CPWF was a 10-year program (2004–2013) designed to address inter-related issues of water scarcity, water productivity, livelihoods, food security, poverty and the environment. Over time, the Program's focus evolved from research on themes to research on development challenges in basins.

As of this writing (November 2013), the program is approaching closure. During the final year, the CPWF reflected on the experience accumulated in 123 projects in ten river basins to develop messages that sum up that experience (see Chapter 3, Table 3.1 for a further breakdown of projects and Appendix for a complete project list). Note that our messages are largely drawn from preceding chapters of this book and from opinion/reflection from Basin Leaders in working with their teams to compile final basin-specific messages.[1] These were shared and discussed in the final peer-assist session with Basin Leaders and management team. They were not usually drawn from the CPWF monitoring and evaluation mechanism. Monitoring and evaluation (M&E) in Phase 1 focused on compliance, not analysis and learning. M&E in Phase 2 added outcome logic models and "most significant change" stories but compilation and quantitative analysis of monitoring data was effectively terminated with the 45 percent cut to the CPWF central program budget (vs 21 percent cut to basin program budgets) during the budget crisis of March 2012.

We summarize the principal messages that the CPWF wishes to convey (Box 9.1). The messages are directed at several audiences, among them researchers interested in development, development workers interested in the contributions of research to problem solving, research managers from national and international institutions, donor and development assistance agencies, NGOs, students and young scientists, the CPWF community, and the CGIAR and its Research Programs. Our intent in sharing these messages is to encourage others to build on the change processes with which the CPWF engaged.

Box 9.1 Six core messages from the CGIAR Challenge Program on Water and Food

Message 1: Water is not scarce, it is the way that it is managed: because of their complexity, addressing water and food issues means tackling wicked problems.

Message 2: Research for development requires dedicated people, time, and continuity to address the wicked problems of water and food.

Message 3: Technical and institutional innovation go together, but innovation is long-term, non-linear and risky.

Message 4: Equitable access to water-related benefits can be achieved through improved water governance, water-related rights and benefit-sharing.

Message 5: Sustainable intensification relies on market incentives and often on water infrastructure.

Message 6: Engagement with decision-makers supported by modeling tools and innovation platforms helps build capacity and consensus, and increases the effectiveness of policy analysis, planning and implementation.

Water is not scarce, it is the way that it is managed: Because of their complexity, addressing water and food issues means tackling wicked problems

[T]he planet has enough water to meet the full range of needs for people and ecosystems in the foreseeable future, but social and environmental equity will only be achieved through [careful] and creative management [of water resources] (Joubert and Trollip, 2012).

Water is not scarce, it is the way that it is managed

Based on the accumulated evidence, the CPWF abandoned its initial assumptions on water scarcity and the need to improve water productivity rapidly. Within a river basin, freshwater can be scarce and abundant at the same time, depending on whether we are talking about blue or green water. Although the dominant demand in water-scarce areas is access, reliable availability of good-quality water is also important (Limpopo). Blue water may be scarce even when green water and rainfall are abundant. Problems of water access and availability and water scarcity often have institutional rather than physical causes, and the relevant institutions will differ depending on a basin's development trajectory (Chapter 2).

Addressing water and food issues means tackling wicked problems

Water and food challenges are often wicked problems that are difficult to solve because their requirements are contradictory, changing, hard to reconcile and not well understood. Solutions require many people to change their mind-sets and behaviors. They require that social and biophysical approaches be integrated using consultative processes to develop collective understanding and responses. Consultative processes and changes in behavior need to be informed by biophysical research at multiple linked levels and focused on the wicked problems at hand (Chapter 3).

Poverty often limits water availability and access but water scarcity does not necessarily cause poverty

Water scarcity was not strongly correlated with economic poverty, which highlights the danger of assuming that improving water availability will reduce poverty. Poverty is more strongly related to the absence of basic services such as safe drinking water, sanitation, health care, education, finance, markets, or farming inputs. How economic poverty hampers water access and availability is more closely associated with the level of development at the community, regional, national and river basin levels (Chapter 2).

The CPWF found that water management is an effective entry point for addressing broader sets of complex and interrelated development objectives. Among the objectives are: producing more food while maintaining the sustainability and resilience of agroecosystems, reducing poverty, ensuring equitable use of resources, preserving ecosystems services and adapting to climate change. Improved water management can help achieve broader development goals as well as being an end in itself (Chapters 1 and 2).

Water productivity is more useful as an entry point to understand limitations to water access and availability than as a principal objective

Success in addressing water-related development challenges usually leads to higher water productivity. The success often results in other improvements as well, for example, land productivity, incomes and ecosystem services. Using water productivity as an indicator of change, however, is different than using it as a principal objective.

Water resources are often poorly understood and not used well. Even in water-rich environments, more efficient use of water resources has huge potential to support agricultural and aquaculture production and livelihoods improvement of farming families and communities (Ganges). In principle, increasing water use efficiency raises the returns to water use (Limpopo). In practice, however, the estimates of water productivity are complex to interpret, especially in systems other than crop production. Nevertheless, water productivity is a useful diagnostic, even where it has limited value as a standalone objective (Chapter 2).

Research for development requires dedicated people, time and continuity to address the wicked problems of water and food

Outputs from research for development can influence decisions that affect outcomes

The CPWF developed a research for development (R4D) model to address relevant development challenges (wicked problems) in six basins (Chapter 4). In R4D, research outputs inform and support decisions leading to development outcomes. Outputs consist of information and insights emerging from research for use in engagement strategies for decision support or negotiation support. These engagement strategies aim to influence knowledge, attitudes and skills of decision-makers at large, and thereby influence outcomes, defined as changes in policy or practice.

The CPWF R4D approach used theory of change (ToC) based on the concepts of complex adaptive systems, learning selection and social networks. ToC encompasses the causes and effects that link research activities to desired changes in policy and practice. It describes the tactics and strategies, including working through partnerships and social networks necessary to achieve desired changes (Chapter 3). R4D uses time frames longer than those of individual projects. R4D is "self-aware," that is, practitioners learn from experience and continually apply this learning to research design, implementation and use.

Using ToC in R4D enables research to focus not only on objectives for change, but how change occurs, with whom, from whose perspective and how it evolves. The objectives can be in terms of knowledge, relationships with people and the environment, or individual and collective actions.

The CPWF conducted R4D on Basin Development Challenges (BDCs) in each of six river basins, identified through a participatory, inclusive process. Addressing BDCs allowed the CPWF to focus on specific relevant problem sets in the context of the broader complex adaptive systems found in basins.

Research can and must inform development processes

The purpose of research on technical and institutional innovation is to provide knowledge. In R4D, knowledge is intended to help inform engagement, dialogue and negotiation. It should be credible and relevant. There are three categories of knowledge from research:

* Knowledge on the nature of the problem or challenge being addressed;
* Knowledge on the design and performance of innovations; and
* Knowledge (ex ante and ex post) on the broader consequences of innovations (Chapter 5).

Scientific research that is credible, relevant and grounded in regular engagement with stakeholders is more likely to produce outputs that can be translated into outcomes (Volta).

Partnerships are the foundation of R4D

The key to successful R4D is a set of diverse, effective, empowered, committed partners who share a common vision of how development challenges can be addressed. The roles played by the CPWF in partnerships included those of convener, engager, negotiator, enabler, space provider and trusted broker—but not executive, boss or chief. Partnership issues include trust, common vision, end-user mandates, strategic alliances, convening power and adaptive management (Chapter 7).

The time-bound nature of the CPWF complicated its engagement in policy processes and its legitimacy to do research and enable change. Decision-making is a long process, vulnerable to multiple influences and driven by personalities. Partners will continue to engage in change processes. Credible and relevant research used by stakeholders in effective R4D increases the likelihood—but does not ensure—that outcomes will be achieved. Through partners, translating outputs to outcomes is likely to continue after the CPWF ends (Chapters 3 and 7).

Partnerships integrate and share diverse elements of local knowledge and innovation processes with other national and international knowledge and experience. Integration and sharing encourage innovation through R4D ("learn by doing and sharing") processes founded in scientific excellence. Innovation is more likely to lead to sustainable outcomes than either local practices alone or promotion of "scientific" technologies from outside the community (Nile).

R4D partnerships should build on existing initiatives, partnerships and networks being mindful of mandates, legitimacy, authority and convening power (Limpopo).

Capacity building is a key outcome of the CPWF

The strengthened capacity of individuals who took part in CPWF is a one of its major outcomes and an important part of its legacy (Chapter 4). Strengthening and transforming Ethiopian institutional and human capacities of all stakeholders was a critical requirement to achieve the full potential benefits of the sustainable land management program (Nile). It is possible to achieve outcomes, but it needs investment in people and partners for a long time. Training young scientists can sow the seeds of future progress in R4D.

The institutional environment for R4D—its leadership, mandates and power dynamics—is a major determinant of its success

The institutional environment within which R4D is conducted determines its success. Interventions to change institutions will fail unless the institutional tendency to preserve the status quo is overcome. A special type of organization is needed to broker, convene, enable and manage the complex relationships among partners, stakeholders and decision-makers (Chapter 4).

The CPWF was designed as a CGIAR reform program with an innovative governance and business model. As structured, however, the host Center retained legal responsibility for the hosted program. The CPWF sought flexibility in implementing its plans while the International Water Management Institute (IWMI) sought to maintain close supervision over CPWF activities. The unstated assumption was that IWMI would be willing to relinquish authority over the CPWF while retaining responsibility for its actions. In the end, this arrangement proved difficult to maintain.

The CPWF was designed with a Phase 2 aimed at moving from outputs to outcomes and a Phase 3 aimed at moving from outcomes to impact, which was termed the "exit strategy." The CGIAR Consortium Board (CB) took the decision to transform this exit strategy into integration with the new CGIAR Research Program on Water, Land and Ecosystems (WLE), probably judging that integration with WLE would yield a similar result as the originally planned exit strategy. The CPWF continued to work with WLE to build on CPWF successes where appropriate to generate outcomes and impact (Chapter 4).

Technical and institutional innovation go together, but innovation is long-term, non-linear and risky

Continuous learning mechanisms can evolve better ways to meet water and food challenges through change that is relevant, appropriate, sustainable and at scale. Farm- and catchment-level technical and institutional innovations have greater impacts when they are widely adopted. Moreover, high-level policy and institutional changes also have greater impacts when they affect large numbers of people over a large area. Some innovations spread spontaneously, but others often need strategies for scaling out. When there is wide-scale adoption of an innovation, however, there may be unexpected positive and negative consequences.

Technical innovation and institutional and policy innovation go hand in hand

In all six basins of Phase 2, the experience from CPWF project teams was that technical innovation and institutional and policy change often go together. One without the other does not usually get to outcomes (Chapter 5). How people collectively manage river systems through adequate institutions and policy often makes the difference between poor and adequate health of the river ecosystem.

This may sound obvious, but it remains counterintuitive to many. The Limpopo team observed that "technology and infrastructure development can be laid out according to timelines and calendars but corresponding institutions are more difficult to establish." R4D consciously aims at integrating mutually reinforcing technical, institutional and policy innovations. Basin teams found that this made some researchers uncomfortable.

Innovation is a long-term, non-linear social process that is risky and unpredictable

Innovation is a long-term, non-linear social process of learning selection that is risky and unpredictable, where success often comes through learning from failure. People experiment by trying novel ways to do things. If they succeed, they may decide to adopt the innovation, adapt it or abandon it. While they experiment they interact with others, who may influence what they decide to do with the innovation. Learning selection applies to institutional as well as technical innovations. The process is spontaneous, although it can be nurtured by facilitators and shepherded by product champions, who play distinct and important roles.

An innovation system is a network of organizations and people, together with the policies that affect their innovative behavior, which bring new processes, new forms of organization and new products into economic use. Innovation itself is often driven by entrepreneurs pursuing market opportunities (Chapters 3 and 5). Because innovation and learning selection are long-term, non-linear social processes, it is important to be flexible and to change plans if the circumstances require it. This is adaptive management. Success takes time, almost always beyond that provided in a 3–5-year project (Limpopo).

Adaptive management allows flexibility

Basin program teams learned the importance of adaptive management for flexibility. The Nile team learned that successful landscape management required reconciling top-down national priorities for soil and water conservation with community needs. The Ganges team revised their research questions several times as they came to understand the interrelationships between components as production systems intensified. The components include the technologies of the farming systems, the coordination and timing of water control, and the design, repair and management of rural infrastructure. Overlaying all was the overlap between national and local government policies and priorities. The Limpopo team built on past achievements but also found new ways to design water infrastructure for multiple uses and to develop markets for small livestock. In the process, partners adjusted to how they perceived and addressed opportunities.

Scaling out innovations involves tailoring to local circumstances

Basin teams agreed that scaling out of innovations involves tailoring to local circumstances. Not all smallholder farmers are the same or have the same resources, goals or responsibilities. Therefore not all technologies fit. Targeting and adapting require attention and investment (Limpopo).

One key to scaling out innovations is to make sure that the innovations are attractive on their own terms. It is necessary to create, align and implement incentives for all parties, and manage risk, to implement sustainable innovative programs at scale (Nile). Moreover, enabling environments are a prerequisite to widespread adoption of technologies (Limpopo).

Several basins used spatial analysis and modeling to guide scaling-out strategies. The Ganges team found that spatial data management tools helped in planning, policy analysis, technology targeting and consensus building. They were able to increase effectiveness of planning, technology targeting, open dialogues and consensus building among multiple stakeholders. They did this by providing access to modeling and spatial analysis at different scales to allow "scenario-based planning and target domain identification."

The Volta team concluded that replicating successful agricultural water management interventions in new locations requires consideration of economic, biophysical, institutional and cultural data. The Targeting Agricultural Water Management Interventions (TAGMI) tool is one way to consider these factors when targeting agricultural water management interventions.

Innovations can have unexpected consequences when used at scale

Market-driven intensification of farm systems and other farm-level innovations may benefit individual farmers in the near term but may not be equitable or sustainable. Innovations may have broader-scale consequences for system resilience, for other water users or for the environment. Research has a role to study and quantify the broader consequences of change and introduce relevant information into engagement strategies (decision support, negotiation support) with decision-makers.

The Nile team noted that promoting single interventions at a mass scale can lead to less than optimal outcomes, and possibly to implementing inappropriate technologies. It is critical, therefore, to distinguish private on-site costs and benefits from downstream or off-site costs and benefits. They suggested that integrated planning and implementation at watershed and basin scales will produce synergies. These can result in important positive impacts on both people's livelihoods and natural resource conservation.

Sustainability of agroecosystems and their related sociological components need to consider the trade-offs between intensifying agriculture and the health of aquatic ecosystems (Volta). Downstream or off-site benefits and costs as well as upstream or on-farm benefits and costs are important considerations. Private investors may need appropriate incentives where the benefits of their investments accrue only as broad public goods. Examples are where the beneficiaries are other stakeholders such as those downstream, or the benefits only accrue after considerable time delay (Nile).

Some basin teams concluded that win-win strategies are possible: benefit-sharing mechanisms (BSMs) help create a virtuous circle between the welfare of people and the ecosystems they live in (Andes). Water, food and the environment can be joined in meaningful ways but they need to be addressed together (Limpopo).

Sometimes there is little information on the scale consequences of change. For example, in the water–poverty–energy nexus in the Mekong, economic feasibility studies fail to account for the true costs of hydropower dams. Moreover, the cumulative impact of hydropower dams remains unknown. Costs of hydropower dams are unevenly distributed but the distribution of costs is difficult to measure. Transparency is a critical ingredient in hydropower planning but there are few incentives for transparency and a lack of enabling information. Finally, there are opportunities to improve the ecological productivity of hydropower reservoirs with modest investments. The broader consequences of these investments, when scaled out, however, are uncertain.

Research can help anticipate and monitor the broader consequences of change

There can be many subtle and unexpected consequences of innovation and change. The challenge for research is to choose which of them are important to anticipate ex ante and which to monitor ex post. These can include changes in land, water and labor productivity; livelihood strategies, including market interactions and labor migration; incomes, poverty and food security; farm household profits and their distribution within the household; climate and market-related risk; human health; gender and equity, including access to resources; cross-scale consequences and externalities; system sustainability and resilience; build-up or loss of social capital; build-up or loss of natural capital; and ecosystem services, biodiversity and environmental quality (Chapter 5).

Equitable access to water-related benefits can be achieved through improved water governance, water-related rights and benefit-sharing

BSMs create a virtuous circle between ecosystems and people's welfare

BSMs are institutional innovations created as a mechanism for sharing water-related costs and benefits among different social groups. They need new cross-sectoral and transboundary institutional arrangements to address high-level issues of water management. These arrangements can enable equitable and sustainable development through improved resource governance and more productive and resilient water management.

The Andes program focused on BSMs. It found that one size does not fit all and that BSMs must be designed within the local social and hydrological context. Indeed, BSMs are a tool for integrated water resource management and adaptation to climate change. They should preferably target watersheds where there is seasonal water supply upstream and a high demand downstream. BSMs are fair and equitable when all stakeholders are provided with all necessary information before they are designed and implemented. The most vulnerable people must develop "hydro-literacy" to avoid power imbalances related to access to information.

BSMs are not only market-driven payments for ecosystem services (PESs); sometimes negotiated, non-market driven BSMs are more suitable. Although they can flourish without supporting regulatory frameworks, well-designed regulations can greatly assist their design and implementation. BSMs should be created as dynamic (rather than static) programs with continuous monitoring and adjustments. Cross-scale feedback mechanisms can inform cycles of dialogue and negotiation for matching high-level priorities with local needs.

A transparent and accountable water governance framework is a prerequisite to improving water management

Most developing countries around the world have water governance frameworks, often imposed by development banks. CPWF experience shows that they are often neither transparent nor accountable, which hampers improvements in water management. The Mekong team observed that "transparency is a critical ingredient to hydropower planning."

The Ganges team, working in polder land in coastal areas, found that "the present institutional coordination is too fragmented and disjointed. There is a need for a transparent and accountable water governance framework for the polders that (i) formalizes and enhances the role of local government institutions in all levels of water governance, and (ii) follows IWRM river-basin governance principles, giving due attention to interactions and interdependence among different scales and sectoral users."

Innovations occur within a policy and political context and the engagement processes involved deal with power relationships (Chapter 7). This is especially true at innovations aimed to improve the transparency and accountability of water governance.

Recognizing the rights of the most vulnerable prevents conflicts and increases water and food security

In basins characterized by strong inequities, vulnerable groups (often women) are caught in poverty traps. CPWF experience showed that recognizing and securing their rights prevents conflicts and increases water and food security.

The Limpopo team observed that, "securing land and water rights for those producers with insecure tenure" was a major institutional change to increase water and food security. For multiple-use water services (MUS), which are often targeted at the poorest populations, they also observed that, "institutional and governance mechanisms are needed for [MUS] because they are complex."

Participatory mechanisms called *Conversatorios* were established in the Andes. These aimed to promote environmental justice, defend territorial rights and social and institutional governance, and help empower the most vulnerable, especially women. "Through enhanced recognition of their rights, women were able to liberate their voice, which led to a transformation of the state-community relationship in the form of increased social accountability and transparency."

Sustainable intensification relies on market incentives and often on water infrastructure

Innovative water management provides many opportunities to raise productivity equitably, improve livelihoods and reduce poverty. Water managers and users are more likely to innovate when they benefit from doing so.

Markets provide incentives for investments in production and resource management

Basin teams found that intensified farm systems deliver improved livelihoods in the near term and were often the foundation for spontaneous innovation and change. Intensification driven by market-value chains changes farm system management. In turn the intensification can drive further adjustments in land and water management.

Some basin teams found huge potential for intensification in irrigated systems: "With advances in crop and aquaculture technologies and available water resources, there is tremendous potential to improve food security and [livelihoods. The technologies include] improved species, varieties, cropping system intensification and diversification on saline soils in coastal zones" (Ganges).

Other basin teams found the same potential in rainfed systems, calling for, "a stronger focus on improving markets, value chains and multi-stakeholder institutions to enhance the benefits and sustainability of rainwater management investments." "Strong value chains in which producers receive a fair share of benefits through appropriate institutions will lead to higher incomes and sustainability of rainwater management interventions" (Nile).

Improved livelihoods through intensification often depend on water infrastructure

In another message we wrote "Water is not scarce, it is the way it is managed." An important part of water management is the design, maintenance, management and governance of water infrastructure.

Summarizing experience from the coastal Ganges, the basin team found that improved livelihoods can emerge from intensification. "Achieving large-scale adoption of innovative production systems and unlocking the potential of water resources requires investment in water management infrastructure." They found that, "Improving drainage is the key intervention and the entry point for cropping intensification and diversification." In the coastal Ganges, water infrastructure includes rural infrastructure not originally intended for water control: "Rural structures (roads, embankments, and culverts) must be considered an integral part of the water management infrastructures. They can effectively form the boundaries of sub-hydrological units, and also units of community water management."

In the Volta, small reservoirs are central to intensification strategies using high-value crops. Two-thirds of the rural population in the Volta live within 3 km of a small reservoir, which directly or indirectly affects their livelihoods. The effects are both positive (food or income security) and negative (waterborne diseases). The Volta team found it striking that young men are least likely to migrate if they live near a small reservoir. Reservoirs directly or indirectly provide employment opportunities and create *de facto* markets for commodities produced in and around the reservoir.

Water infrastructure can benefit some groups but harm others. Hydropower dams in the Mekong produce electricity that benefits urban areas throughout Southeast Asia, but harms fishers and farmers living downstream—often far downstream—from the dams.

Water infrastructure depends on adequate maintenance as well as good design

Basin teams found that infrastructure is most effective in fostering intensification when it receives routine maintenance.

The Limpopo team observed that, "Small reservoirs typically fail for many reasons but lack of regular maintenance is the most evident . . . [D]evelopment, installation or rehabilitation of water infrastructure should be done with a multiple-use approach." The Volta team also found numerous problems with small reservoir maintenance.

The Ganges team focused on water infrastructure maintenance as one of several complementary factors. They found that livelihoods can be improved through intensified farm systems but that intensification depends on better water control, especially drainage. Drainage depends on improved maintenance, which in turn depends on institutional and governance change.

> [Deferred] maintenance of infrastructures is the Achilles heel of water management in the poldered coastal zone. Maintenance is often deferred due to lack of incentives and funds. These can be solved through a three-tier strategy: a) Community level: Improving income of the water management organizations by increasing contributions from the community; b) local government level: effective use of local government social safety net funds in maintenance of infrastructure; and c) central government and donors: creating a self-replenishing Trust Fund.
>
> (CPWF, 2013)

Design of hydropower dams was a major issue in the Mekong because good design can generate additional benefits for more groups. It can also make cost sharing equitable while maintaining power generation. "Costs of hydropower dams are unevenly distributed . . . [but] hydropower can be multi-purpose—relatively simple strategies can achieve this."

Water infrastructure design and maintenance depend on institutions, governance and policies

Basin teams found that successful design and maintenance of water infrastructure often depends on governance arrangements and policies that support and facilitate the corresponding efforts of communities and local institutions. The Limpopo team observed that, "because [MUS] are complex, institutional and governance mechanisms are needed (to enable their proper design and maintenance)."

In the Mekong basin, the cumulative impact of hydropower dams (designed and operated in uncoordinated "cascades") and other water infrastructure remains unknown by local communities and institutions. They are often absent from discussions of the design and maintenance of water infrastructure.

Engagement with decision-makers, supported by modeling tools and innovation platforms, helps build capacity and consensus, and increases the effectiveness of policy analysis, planning and implementation

Groups with differing interests in water development, including interests in social and environmental equity, can negotiate ways to share pertinent benefits and costs so that everyone is better off. Sustained and inclusive engagement with decision-makers and communities, at different scales and in different contexts, at all levels, informed by credible and relevant research products, helps research contribute to positive development outcomes.

Engagement platforms are useful for finding innovative solutions to complex problems affecting diverse groups with differing interests. They enable people to work together, connect across different levels and identify workable solutions.

Outcomes emerge from engagement strategies informed by research

The CPWF has a history of using research outputs to inform strategies to engage with decision-makers. Successful engagement, however, needs to extend beyond the lifetime of an individual project. "Facilitating engagement to assess options and opportunities enables farmers to identify and choose the best options for themselves"; however, these processes take time (Limpopo).

R4D redefines who decision-makers are. Decision-makers are normally assumed to be senior government officials. We say assumed because despite the many references to decision-makers in the literature, they are seldom named or identified. Within the context of R4D, however, everyone is a decision-maker. Decisions are made by groups and individuals, among them senior government officials, business owners, trade unionists, NGO staff, members of farmers' associations and scientists.

Innovation platforms are an effective mechanism for engagement

Platforms are variously known as engagement platforms, multi-stakeholder platforms or innovation platforms. In general terms, an engagement platform is an opportunity for individuals and people representing organizations with different backgrounds and interests to come together to diagnose problems, identify opportunities and implement solutions (Chapter 7).

Innovation platforms provide spaces for a wide range of stakeholders to exchange knowledge, learn, and develop joint initiatives to solve agricultural development challenges. Successful innovation can only happen when stake-holders have an interest in working together to acquire knowledge and find solutions. The research community cannot bring about innovation on its own (Volta). They facilitate engagement to assess options and opportunities and enable farmers to identify and choose the best options for themselves. They help reduce the cost of searching for and reaching markets (Limpopo).

1   The most successful engagement platforms are self-reliant, demand driven, evolve over time and embrace multiple perspectives.
2   They build on what is already there rather than set up new platforms and systems.
3   Engagement platforms are not neutral mechanisms. They aim to promote change so they are disruptive by nature. Power relationships influence the course of dialogue and negotiation.
4   Engagement platforms can be useful vehicles for exploring strategies to boost productivity, improve natural resources management, strengthen value chains and adapt to climate change.
5   Establishing engagement platforms at several levels is one way to stimulate vertical and horizontal coordination for greater impact.
6   Platforms can empower local actors to hold higher levels of government to account.
7   Markets provide clear incentives for investments in production.

The Volta team observed that, "successful integrated water resources man-agement depends on interactions between multiple actors at different scales, which are often beyond everyday considerations." They also noted that "the companion modeling approach is a good framework to highlight interactions between actors and allows for a collective decision-making process to unfold."

The Mekong team also noted that "Multi-stakeholder platforms are a viable decision-making mechanism on hydropower issues," while the Andes team observed that "Fair and equitable benefit-sharing mechanisms are designed and implemented when all stakeholders are provided with all the necessary information."

Community empowerment helps achieve long-term benefits

In the Nile and in the Andes, the CPWF experienced the importance of local community empowerment and leadership, especially where collective action was needed. Based on demand, equity and inclusiveness, it was critically important to achieve long-term benefits and sustainable outcomes from rainwater management and benefit-sharing programs.

Empowerment may also require regulations and safeguards, as noted by the Mekong team: "Protocols and safeguards are critical to implementing and monitoring hydropower dams." Successful integrated water resources management depends on interactions among multiple actors at different scales. The companion modeling approach was a good framework to highlight interactions among actors and allow for a collective decision-making process to unfold (Volta). Processes to facilitate engagement for assessing options and opportunities enabled farmers to identify and choose the best options for themselves (Limpopo).

Some kinds of engagement platforms can empower local actors to hold higher levels of government to account (Andes). Power relationships influenced the course of dialogue and negotiation. They figured prominently in the process of moving from outputs to outcomes (Chapter 7).

Modeling tools developed with stakeholders can support and inform engagement

Modeling tools developed with stakeholders supported capacity and consensus building and increased the effectiveness of policy analysis, planning and implementation. The CPWF developed and applied various modeling tools and invested in systematic data acquisition, especially spatially explicit data at basin and lower scales using geographic information systems.

The Ganges team observed that, "the effectiveness of planning, technology targeting, open dialogues and consensus building among multiple stakeholders can be increased by access to modeling and spatial analysis at different scales for scenario-based planning and target domain identification."

The Nile team observed that, "adapting and using the growing suite of new models and learning and planning tools, including those piloted by the Nile BDC, combined with stronger learning processes, increased the effectiveness of planning, implementation, and capacity building." These models and tools included: (i) integrated hydrologic, water resource and economic models for planning, scaling out and impact assessments; (ii) user-friendly tools to facilitate local-level learning, training, and identifying appropriate interventions; and (iii) a centralized database for geographical and other data, which could enhance the efficiency of planning, implementation, learning and evaluation processes.

In the Mekong the CPWF observed that only models developed with substantial participation of stakeholders through dialogues or participatory modeling gained meaning for researchers and decision-makers.

Conclusions

In this final chapter we summarized our program messages for the purpose of encouraging others to build on the change processes with which we have engaged. By doing this we have two ambitions:

The ambition to inform a new R4D agenda

CPWF developed a model of research for development that we feel is viable. It requires dedicated people, time and continuity. In the CPWF's six basins of Phase 2, the most compelling outcomes were achieved when researchers had engaged over 5–10 years with key stakeholders and related change processes, whether as part of or in concert with CPWF activities.

In this model, outputs from R4D aim to influence decisions that affect development outcomes. Research can and must inform development processes. Consequently, the institutional environment for R4D—its leadership, mandates and power dynamics—is a major determinant of its success. The usefulness of this model is illustrated by the many examples of outcomes described in this book. Numerous partners and researchers are choosing to use the CPWF R4D model. The continued momentum of R4D in basins, building on CPWF achievements, is being carried forward thanks to the resilient partnerships and engagement processes the CPWF put in place.

The ambition to influence donor investment patterns

Because of the necessary time to contribute to development outcomes, we recommend that donors consider using time frames for investment that go beyond the normal 3–5 years. This does not preclude financing in tranches where progress on R4D along impact pathways can be confirmed. But long-term financing on other major issues such as climate change (more than 20 years) has obviously borne fruit. So, why not on water and food which are recognized as major challenges for the 21st century?

We also recommend that donors recognize that grand schemes that implicitly assume that "one size fits all" can be misguided and can carry a number of dangers. A larger number of smaller-scale investments in locally adapted solutions to wicked problems can ultimately have more substantial development impacts on larger numbers of people. Interventions and investments in R4D at the level of landscapes, catchments, sub-basins and basins can be very effective with large cumulative impacts.

Acknowledgments

The research and work that is described in this book was carried out through the CPWF, which was funded by a number of agencies throughout its ten-year life span. These partnerships went beyond funding and included mutual learning, capitalization and cross-fertilization of projects. We thank the following donor partners for their generous support and time: UK Department for International Development, the European Commission, the International Fund for Agricultural Development, the Swiss Agency for Development and Cooperation, the Australian Agency for International Development, the German Gesellschaft für Internationale Zusammenarbeit, the Swedish International Development Cooperation Agency, the World Bank, and the individual countries, France, the Netherlands, Norway, Sweden, Denmark and New Zealand.

Notes

1   Basin messages referred to in this chapter are at present only available in unpublished internal correspondence.

References

CPWF (2013) *Messages from the CGIAR Challenge Program on Water and Food Ganges Basin Development Challenge (GBDC): Rethinking water management to unlock the production potential of the polders of the coastal zone of Bangladesh*, CPWF Working Paper, CGIAR Challenge Progam on Water and Food, Colombo, Sri Lanka, ganges-bdc.wikispaces.com/file/view/GBDC+Messages+bulletin.pdf (accessed 20 April 2014).

Joubert, M. and Trollip, K. (2012) *Streams of innovation: Improving people's lives through research on water and food*, CGIAR Challenge Program on Water and Food, Colombo, Sri Lanka, results.waterandfood.org/bitstream/handle/10568/16976/1_IFWF_Streams%20of%20innovationDocLR.pdf?sequence=1 (accessed 21 April 2014).

Appendix

Projects financed by the CGIAR Challenge Program on Water and Food

Project number	Project title	Lead Institution	Dates	Basin(s)
PN 1	Increased food security and income in the Limpopo Basin through integrated crop, water and soil fertility options and public–private partnerships.	International Crop Research Institute for the Semi-arid Tropics (ICRISAT), Malawi	Start: 1-Jan-05 End: 31-Dec-09	Limpopo
PN 2	Water productivity improvement of cereals and foods legumes in the Atbara Basin of Eritrea.	International Center for Agricultural Research in the Dry Areas (ICARDA), Syria	Start: 15-Jul-04 End: 14-Jul-09 Extension: 30-Apr-10	Nile
PN 5	Enhancing rainwater and nutrient use efficiency for improved crop productivity, farm income and rural livelihoods in the Volta Basin.	International Crop Research Institute for the Semi-arid Tropics (ICRISAT), Niger	Start: 15-Jun-04 End: 14-Jun 09 Extension: 14-Dec-09	Volta
PN 6	Empowering farming communities in Northern Ghana with strategic innovations and productive resources in dry land farming.	Savanna Agricultural Research Institute (SARI), Ghana	Start: 15-Jun-04 End: 14-Jun-09	Volta
PN 7	Development of technologies to harness the productivity potential of salt-affected areas of the Indo-Gangetic, Mekong and Nile River basins.	International Rice Research Institute (IRRI), Philippines	Start: 15-Jun-04 End: 14-Jun-08 Extension: 31-Dec-08	Indus–Ganges 60%, Mekong 40%

Project number	Project title	Lead Institution	Dates	Basin(s)
PN 8	Improving on-farm agricultural water productivity in the Karkheh River basin.	International Center for Agricultural Research in the Dry Areas (ICARDA), Syria	Start: 01–Sep–04 End: 31–Aug–08 Extension: 1st 31–Dec–08, 2nd 30–Jun–09	Karkheh
PN 10	Managing water and land resources for sustainable livelihoods at the interface between fresh and saline water environments in Vietnam and Bangladesh.	International Rice Research Institute (IRRI), Philippines	Start: 01–Jun–04 End: 31–Dec–07 Extension: 30–Jun–08	Indus–Ganges 40%, Mekong 60%
PN 11	Rice landscape management for raising water productivity, conserving resources, and improving livelihoods in upper catchments of the Mekong and Red River basins.	International Rice Research Institute (IRRI), Philippines	Start: 01–Nov–05 End: 31–Oct–09 Extension: 30–Apr–10	Mekong 60%, Other 40% (Red River basin)
PN 12	Conservation agriculture for the dry-land areas of the Yellow River basin: Increasing the productivity, sustainability, equity and water use efficiency of dry-land agriculture, while protecting downstream water users.	Centro Internacional de Mejoramiento de Maíz y Trigo (CIMMYT), Mexico	Start: 01–Mar–05 End: 28–Feb–09 Extension: 30–Oct–09	Yellow
PN 15	Quesungual slash and mulch agroforestry system (QSMAS): Improving crop water productivity, food security and resource quality in the sub-humid tropics.	Centro Internacional de Agricultura Tropical (CIAT), Colombia	Start: 01–Sep–04 End: 31–Aug–07 Extension: 1st 31–Aug–08, 2nd 15–Dec–08, 3rd 15–Mar–09	Lempira (Honduras) 60%, Calico (Nicaragua) 25%, Andes (Colombia) 15%
PN 16	Developing a system of temperate and tropical aerobic rice in Asia (STAR).	International Rice Research Institute (IRRI), Philippines	Start: 01–Oct–04 End: 30–Sep–07 Extension: 31–Mar–08	Indus–Ganges 25%, Mekong 25%, Yellow River 25%, Other 25%

PN	Project title	Lead institution	Dates	Basin
PN 17	The challenge of integrated water resource management for improved rural livelihoods: Managing risk, mitigating drought and improving water productivity in the water scarce Limpopo Basin.	WaterNet, Zimbabwe	Start: 01-Sep-04 End: 31-Aug-08 Extension: 31-Aug-09	Limpopo
PN 19	Improved water and land management in the Ethiopian highlands and its impact on downstream stakeholders dependent on the Blue Nile.	International Water Management Institute (IWMI), Ethiopia	Start: 01-May-07 End: 30-Sep-09 Extension: 1st 31-Dec-09, 2nd 30-Apr-10	Nile
PN 20	Sustaining inclusive collective action that links across economic and ecological scales in upper watersheds (SCALES).	Centro Internacional de Agricultura Tropical (CIAT), Colombia	Start: 01-Sep-04 End: 30-Jun-08	Andes 50%, Nile 50%
PN 22	Payment for environmental services as a mechanism for promoting rural development in the upper watersheds of the tropics.	Consortio para del Desarrollo Sostenible de la Ecoregión Andina (CONDESAN), Ecuador	Start: 15-Jan-05 End: 14-Jan-08 Extension: 1st 15-Jun-08, 2nd 15-Dec-08	Andes 50%, Nile 50%
PN 23	Linking community-based water and forest management for sustainable livelihoods of the poor in fragile upper catchments of the Indus–Ganges Basin.	International Water Management Institute (IWMI), Nepal	Start: 01-Apr-05 End: 31-Mar-08 Extension: 30-Sep-08	Indus–Ganges
PN 24	Strengthening livelihood resilience in upper catchments of dry areas by integrated natural resources management.	International Center for Agricultural Research in the Dry Areas (ICARDA), Syria	Start: 01-Sep-04 End: 31-Aug-08 Extension: 1st 31-Dec-08, 2nd 30-Jun-09	Karkheh
PN 25	Companion modeling for resilient water management: Stakeholders' perceptions of water dynamics and collective learning at the catchment scale.	Centre de coopération internationale en recherche agronomique pour le développement (CIRAD), France	Start: 01-Jan-06 End: 31-Dec-08 Extension: 31-Dec-09	Mekong

Project number	Project title	Lead Institution	Dates	Basin(s)
PN 28	Multiple-use water services (MUS).	International Water Management Institute (IWMI), Sri Lanka	Start: 15-Jun-04 End: 14-Jun-08 Extension: 01-May-09	Andes 25%, Indus–Ganges 20%, Limpopo 25%, Mekong 20%, Nile 10%
PN 30	Wetlands-based livelihoods in the Limpopo Basin: Balancing social welfare and environmental security.	International Water Management Institute (IWMI), South Africa	Start: 15-Jun-04 End: 14-Dec-07 Extension: 1st 15-Jun-08, 2nd 31-Aug-08	Limpopo
PN 34	Improved fisheries productivity and management in tropical reservoirs.	WorldFish Center (WFC), Egypt	Start: 01-Mar-05 End: 31-Aug-08 Extension: 31-May-09	Indus–Ganges 33%, Nile 33%, Volta 33%
PN 35	Community-based fish culture in irrigation systems and seasonal floodplains.	WorldFish Center (WFC), Malaysia	Start: 01-Apr-05 End: 30-Mar-10	Indus–Ganges 55%, Mekong 35%, Niger 10%
PN 36	Improved planning of large dam operation: Using decision support systems to optimize livelihood benefits, safeguard health and protect the environment.	International Water Management Institute (IWMI), Ethiopia	Start: 01-Jan-05 End: 31-Dec-08 Extension: 1st 30-Jun-09, 2nd 31-Dec-09	Nile
PN 37	Nile Basin livestock water productivity.	International Livestock Research Institute (ILRI), Ethiopia	Start: 15-Jun-04 End: 14-Jun-08 Extension: 30-Jun-09	Nile
PN 38	Safeguarding public health concerns, livelihoods and productivity in wastewater irrigated urban and peri-urban vegetable farming.	Kwame Nkrumah University of Science and Technology (KNUST), Ghana	Start: 01-Dec-04 End: 30-Nov-07 Extension: 30-Jun-08	Volta

ID	Title	Institution	Dates	River basins
PN 40	Integrating knowledge from computational modeling with multi-stakeholder governance: Towards more secure livelihoods through improved tools for integrated river basin management.	International Food and Policy Research Institute (IFPRI), USA	Start: 15-Jun-04 End: 14-Jun-08 Extension: 31-Dec-08	Andes 50%, Volta 50%
PN 42	The international training and research program on groundwater governance in Asia: Theory and practice.	International Water Management Institute (IWMI), India	Start: 01-Sep-05 End: 30-Nov-08 Extension: 01-Mar-09	Indus-Ganges 50%, Yellow 50%
PN 46	Small multi-purpose reservoir ensemble planning.	International Water Management Institute (IWMI), Ghana	Start: 15-Jun-04 End: 14-Jun-07 Extension: 1st 31-Jul-08, 2nd 31-Dec-08	Limpopo 33%, São Francisco 33%, Volta 33%
PN 47	African models of trans-boundary governance.	International Water Management Institute (IWMI), South Africa	Start: 15-Jun-04 End: 14-Jun-07 Extension: 14-Dec-07	Limpopo 50%, Volta 50%
PN 48	Strategic analysis of India's river linking project.	International Water Management Institute (IWMI), India	Start: 01-Apr-05 End: 28-Feb-08 Extension: 1st 15-Dec-08, 2nd 31-Mar-09, 3rd 15-May-09	Indus-Ganges
PN 50	Enhancing multi-scale Mekong water governance.	Chiang Mai University (CMU), Thailand	Start: 01-Jan-06 End: 31-Dec-09 Extension: 30-Apr-10	Mekong
PN 51	Wastewater irrigation opportunities and risks.	University of Copenhagen, Denmark	Start: 01-Jan-05 End: 31-Dec-06	Volta
PN 52	Mekong fisheries management institutions.	Institute of Fisheries Management and Coastal Community Development, Denmark	Start: 01-Feb-05 End: 31-Jan-06	Mekong

Project number	Project title	Lead Institution	Dates	Basin(s)
PN 53	Food and water security under global change: Developing adaptive capacity with a focus on rural Africa.	International Food and Policy Research Institute (IFPRI), USA	Start: 05–Mar–07 End: 31–Jan–09	Nile 50%, Limpopo 50%
PN54a	Basin Focal Project coordination and impact assessment.	Challenge Program on Water and Food (CPWF), Sri Lanka	End: 30–Jun–09	Across basins
PN54b	Basin Focal Project cost benefit analysis.	Challenge Program on Water and Food (CPWF), Sri Lanka	End: 31–Dec–08	Across basins
PN 55	Basin Focal Project, Volta.	L'institut de recherche pour le développement (IRD), France	Start: 01–Sep–05 End: 28–Feb–08 Extension: 1st 31–Oct–08, 2nd 31–Mar–09, 3rd 31–Dec–09	Volta
PN 56	Basin Focal Project, São Francisco.	University of California Davis (UCD), USA	Start: 01–Sep–05 End: 28–Feb–08 Extension: 31–Dec–08	São Francisco
PN 57	Basin Focal Project, Karkheh.	International Water Management Institute (IWMI)	Start: 01–Sep–05 End: 28–Feb–08 Extension: 31–Dec–08	Karkheh
PN 58	Basin Focal Project, Mekong.	Commonwealth Scientific and Industrial Research Organisation (CSIRO), Australia	Start: 01–Sep–05 End: 28–Feb–08 Extension: 1st 30–Jun–08, 2nd 30–Sep–08	Mekong

PN	Project	Institution	Dates	Basin
PN 59	Basin Focal Project, Nile.	International Water Management Institute (IWMI), Ethiopia	Start: 01-Jan-08 End: 31-Dec-09 Extension: 30-Apr-10	Nile
PN 60	Basin Focal Project, Indus Ganges.	International Water Management Institute (IWMI), India	Start: 01-Jan-08 End: 31-Dec-09	Indus-Ganges
PN 61	Basin Focal Project, Yellow River.	International Food and Policy Research Institute (IFPRI), USA	Start: 01-Dec-07 End: 30-Nov-09 Extension: 30-Jun-10	Yellow
PN 62	Basin Focal Project, Limpopo.	Agricultural Research Center (ARC)/ Food, Agriculture and Natural Resources Policy Analysis Network (FANRPAN), South Africa	Start: 08-Jan-08 End: 07-Jan-10 Extension: 1st 30-Apr-10, 2nd 30-Jun-10	Limpopo
PN 63	Basin Focal Project, Andes.	Kings College, London (KCL), UK	Start: 01-Dec-07 End: 30-Nov-09 Extension: 31-Mar-10	Andes
PN 64	Basin Focal Project, Niger.	L'institut de recherche pour le développement (IRD), France	Start: 01-Jan-08 End: 31-Dec-09 Extension: 30-Apr-10	Niger
PN 65	Shallow groundwater irrigation in the White Volta Basin.	International Water Management Institute (IWMI), Ghana	Start: 01-Dec-07 End: 30-Apr-09 Extension: 30-Apr-10	Volta
PN 66	Water rights in informal economies in the Limpopo and Volta Basins.	International Water Management Institute (IWMI), South Africa	Start: 01-Dec-07 End: 31-Jul-09 Extension: 15-Dec-09	Limpopo

Project number	Project title	Lead Institution	Dates	Basin(s)
PN 67	Improving Mekong water resources investment and allocation choices.	Griffin Natural Resource Management (NRM), Australia	Start: 01–Jan–08 End: 31–Dec–09 Extension: 31–Mar 10	Mekong
PN 68	Improving water productivity, reducing poverty and enhancing equity in mixed crop–livestock systems in the Indo–Gangetic Basin.	International Water Management Institute (IWMI), India	Start: 01–Mar–08 End: 28–Feb–10 Extension: 30–Apr–10	Indus–Ganges
PN 69	Valuing the role of living aquatic resources to rural livelihoods in multiple–use, seasonally inundated wetlands in the Yellow River basin of China, for improved governance.	WorldFish Center (WFC), Malaysia	Start: 01–Jan–08 End: 31–Dec–09 Extension: 31–Mar–10	Yellow River
PN 70	Socioeconomic and technical considerations to mitigate land and water degradation in the Peruvian Andes.	International Food and Policy Research Institute (IFPRI), USA	Start: 01–Apr–08 End: 31–Mar–10	Andes
PN 71	Commune agroecosystem analysis to support decision making for water allocation for fisheries and agriculture in the Tonle Sap wetland system.	International Water Management Institute (IWMI), Sri Lanka	Start: 25–Apr–08 End: 25–Apr–10 Extension: 25–Jun–10	Mekong
PN 72	Improving resilience in small–scale fisheries.	WorldFish Center (WFC), Malaysia	Start: 01–Jan–08 End: 31–Dec–09	Niger
SG 501	Participatory water resource planning and development for economic sufficiency through learning alliance approach in the northeast of Thailand.	Khon Kaen University (KKU), Thailand	Start: 15–Dec–05 End: 14–Dec–06	Mekong
SG 502	Demonstration and documentation of innovative market–based strategies to realize agricultural income through increased on–farm water productivity and market integration.	International Development Enterprises (IDE), Cambodia	Start: 01–Feb–06 End: 31–Jul–07	Mekong

ID	Description	Organization	Dates	Basin
SG 503	Conditions for sustainable adoption of water and moisture system innovations in Nile River basin: Case of Makanya watershed in Tanzania.	Soil-Water Management Research Group (SWMRG), Tanzania	Start: 02-Jan-06 End: 30-Jun-07	Nile
SG 504	Increasing water use efficiency in rice using principles of System of Rice Intensification (SRI) and green mulch in Northeast Thailand.	Asian Institute of Technology (AIT), Thailand	Start: 01-Jan-06 End: 30-Jun-07	Mekong
SG 505	Katalysis: Enabling endogenous potential for improved management and conservation of water resources in semi-arid Andean ecosystems.	World Neighbors, Andes Area Program (WN-AAP), Ecuador	Start: 15-Jan-06 End: 14-Jul-07	Andes
SG 506	Improving catchment and use efficiency of water for high-value dry season crops.	Savanna Agricultural Research Institute (SARI), Ghana	Start: 01-Feb-06 End: 31-Jul-07	Volta
SG 507	Development and testing of training materials and information for scaling up dissemination of micro-irrigation and associated water-control technologies designed for small plot systems.	International Development Enterprises (IDE), Bangladesh	Start: 01-Jan-06 End: 30-Jun-07	Indus-Ganges
SG 508	Selecting and scaling up water-efficient farming and groundwater recharge systems among 3,000 small-scale farmers	International Development Enterprises (IDE), Bangladesh	Start: 02-Jan-06 End: 30-Jun-07	Indus-Ganges
SG 509	Sustainable water management for food security for smallholder farming communities in Tigray, northern Ethiopia.	Institute for Sustainable Development (ISD), Ethiopia	Start: 01-Jan-06 End: 30-Jun-07	Nile
SG 510	Associated cropping and enhanced rainwater harvesting to improve food security and sustainable livelihoods of peasant farmer associations.	Fundación de Expresión Intercultural, Educativa y Ambiental (FUNDAEXPRESIÓN), Colombia	Start: 01-Jan-06 End: 30-Jun-07	Andes

Project number	Project title	Lead Institution	Dates	Basin(s)
SG 511	Outscaling and upscaling community-based water management strategies in the Karkheh River basin.	Centre for Sustainable Development (CENESTA), Iran	Start: 01–Feb–06 End: 31–Jul–07	Karkheh
SG 512	Test marketing low cost irrigation solutions in Bihar and Jharkhand: Irrigation solution matrix for smallholder farmers.	International Development Enterprises (IDE), India	Start: 01–Jan–06 End: 30–Jun–07	Indus–Ganges
SG 513	Food security in Southern Uganda.	St. Jude Family Projects and Organic Training Centre (SJFPOTC), Uganda	Start: 01–Jan–06 End: 30–Jun–07	Nile
SG 514	Sekororo rainwater harvesting project.	World Vision (WV), South Africa	Start: 01–Jan–06 End: 30–Jun–07	Limpopo
Andes 1	On designing and implementing benefit-sharing mechanisms.	Centro Latinoamericano para el Desarrollo Rural (RIMISP), Ecuador	Start: 01–Jun–10 End: 31 –Dec–13	Andes
Andes 2	Assessing and anticipating the consequences of introducing benefit-sharing mechanisms.	Centro Internacional de Agricultura Tropical (CIAT), Colombia	Start: 01–Jan–10 End: 30–Nov–13	Andes
Andes 3	On designing and implementing benefit-sharing mechanisms.	King's College London (KCL), UK	Start: 01–Dec–10 End: 31–Dec–13	Andes
Andes 4	Coordination and multi-stakeholder platforms (coordination and change project).	Consortio para del Desarrollo Sostenible de la Ecoregión Andina (CONDESAN), Ecuador	Start: 01–Mar–10 End: 31–Dec–13	Andes

Project	Institution	Dates	Basin
Ganges 1 Resource profiles, extrapolation domains, and land-use patterns.	International Rice Research Institute (IRRI), Philippines	Start: 30-April-11 End: 30-April-14 Extension: 31-Dec-14	Ganges
Ganges 2 Productive, profitable and resilient agriculture and aquaculture systems.	International Rice Research Institute (IRRI), Philippines	Start: 30-April-11 End: 30-April-14 Extension: 31-Dec-14	Ganges
Ganges 3 Water governance and community-based management.	International Water Management Institute (IWMI), Bangladesh	Start: 30-April-11 End: 30-April-14	Ganges
Ganges 4 Assessment of the impact of anticipated external drivers of change on water resources of the coastal zone.	Institute of Water Modeling (IWM), Bangladesh	Start: 30-April-11 End: 30-April-14 Extension: 31-Dec-14	Ganges
Ganges 5 Coordination and change-enabling project.	WorldFish Center (WFC), Malaysia	Start: 30-April-11 End: 30-April-14 Extension: 31-Dec-14	Ganges
Limpopo 1 Targeting and scaling out.	Stockholm Environment Institute (SEI), Sweden	Start: 01-Oct-10 End: 30-Sep-13	Limpopo
Limpopo 2 Small-scale water infrastructure.	Agricultural Research Center (ARC), South Africa	Start: 01-Oct-10 End: 31-Dec-13 Extension: 21-Feb-14	Limpopo
Limpopo 3 Farm systems and risk management.	International Crop Research Institute for the Semi-arid Tropics (ICRISAT), South Africa	Start: 01-Aug-10 End: 30-Nov-13	Limpopo
Limpopo 4 Water Governance.	WaterNet, Zimbabwe	Start: 01-Jan-11 End: 31-Dec-13	Limpopo

Project number	Project title	Lead Institution	Dates	Basin(s)
Limpopo 5	Learning for innovation and adaptive management (coordination and change project).	Food, Agriculture and Natural Resources Policy Analysis Network (FANRPAN), South Africa	Start: 01-Oct-10 End: 30-Sep-13	Limpopo
Mekong 1	Optimizing reservoir management for livelihoods.	International Water Management Institute (IWMI), Sri Lanka	Start: 01 April-10 End: 31-Dec-13 Extension: 09-Mar-14	Mekong
Mekong 2	Water valuation.	WorldFish Center (WFC), Malaysia	Start: 01-Jun-10 End: 31-Aug-13 Extension: 31-Dec-13	Mekong
Mekong 3	Optimizing the management of cascades or systems of reservoirs at catchment level.	International Center for Environmental Management (ICEM), Vietnam	Start: 01-April-10 End: 30-Sept-13 Extension: 14-Feb-14	Mekong
Mekong 4	Water governance—water storage infrastructure.	Asian Institute of Technology (AIT), Thailand	Start: 01-Mar-10 End: 31-Dec-13 Extension:31-July-13	Mekong
Mekong 5	Coordination of multi-stakeholder platforms (coordination and change project).	Challenge Program on Water and Food (CPWF), Lao PDR	Start: 1-Nov-10 End: 31-Dec-13 Extension: 30-Apr-2014	Mekong
Mekong 6	Hydropower governance and multi-stakeholder platforms.	Challenge Program on Water and Food (CPWF), Lao PDR / Mekong Program on Water,	Start: 21-Dec-2009 End: 30 Sep-2010 Extension: 31-May-2012	Mekong

		Environment and Resilience (M-POWER), Lao PDR		
Mekong 7	M-POWER– CPWF Research Fellowship Program.	Asian Institute of Technology (AIT), Thailand	Start: 01-Aug-11 End: 31-Dec-13 Extension: 28-Feb-2014	Mekong
Mekong 8	Improving hydropower decision-making processes in the Mekong Basin.	D-Foundation, Thailand	Start: 01-May-12 End: 31-Dec-13 Extension: 31-Jan-2014	Mekong
Mekong 9	Improving Mekong dam dialogues: A participatory assessment of dams and livelihoods of the Mekong.	Participatory Development Training Centre (PDTC), Lao PDR	Start: 15-Oct-12 End: 31-Dec-13 Extension: 15-Feb-2014	Mekong
Mekong 10	Knowledge and institutional systems in the management and coordination of hydropower social safeguards: Hydropower development in Attapeu Province.	National University of Laos (NUAL), Lao PDR	Start: 15-Oct-12 End: 31-Dec-13 Extension: 31-Jan-2014	Mekong
Mekong 11	Bridging the hydropower policy-implementation gap: Communications and feedback mechanisms to improve participation in decision-making for local land and water use (BPIG).	Village Focus International (VFI), Lao PDR	Start: 01-Oct-12 End: 31-Dec-13 Extension: 3-Jan-2014	Mekong
Mekong 12	The impact of water supply infrastructure (WSI) on floods and drought in the Mekong region and the implications for food production.	International Center for Environmental Management (ICEM), Vietnam	Start: 01-Oct-12 End: 31-Dec-13 Extension: 28-Feb-2014	Mekong
Mekong 13	Balancing the scales: Gender justice in hydropower.	Oxfam, Australia	Start: 15-Oct-12 End: 31-Dec-13 Extension: 28-Feb-2014	Mekong

Project number	Project title	Lead Institution	Dates	Basin(s)
Mekong 14	Potential for increasing the role of renewables in Mekong power supply.	International Center for Environmental Management (ICEM), Vietnam	Start: 15–Oct-12 End: 31-Dec-13 Extension: 28-Feb-2014	Mekong
Mekong 15	Optimising fish-friendly criteria into the design of mini-hydro schemes in the Lower Mekong Basin.	National University of Laos (NUAL), Lao PDR	Start: 15–Oct-12 End: 31-Dec-13 Extension: 31-Jan-2014	Mekong
Mekong 16	Fostering evidence–based IWRM in the Stung Pursat catchment (Tonle Sap Great Lake) Cambodia.	Hatfield Consultants Group, Canada	Start: 15–Oct-12 End: 31-Dec-13	Mekong
Mekong 17	The impact of reservoirs on water resources and socio-economy: A comprehensive study of the Sre Pok River (Vietnam).	Institute of Water Resources Planning (IWRP), Vietnam	Start: 01–Mar –13 End: 31-Dec-13 Extension: 30-Jan-2014	Mekong
Mekong 18	Experimenting with a bottom–up multi-stakeholder platform supported by modelling games in the NT–NK basin.	Centre de coopération internationale en recherche agronomique pour le développement, (CIRAD), France	Start: 01-Jun-13 End: 31-Dec-13 Extension: 15-Jan-2014	Mekong
Mekong 19	Fisheries and aquaculture production in reservoirs in Lao PDR.	Mekong Development Center (MDC), Lao PDR	Start: 29–Mar-13 End: 31-Dec-13 Extension: 15-Feb-2014	Mekong
Nile 1	Learning from the past.	Consultancy project	Start: 10-Mar-10 End: 10-Sept-10	Nile
Nile 2	Integrated rainwater management strategies—technologies, institutions and policies.	International Water Management Institute (IWMI), Ethiopia	Start: 01–Mar-10 End: 31-Dec-13 Extension: 09-Mar-2014	Nile

Nile 3	Targeting and scaling out.	International Livestock Research Institute (ILRI), Ethiopia	Start: 01–Mar–10 End: 28-Feb-12 Extension: 31-Dec-12	Nile
Nile 4	Assessing and anticipating consequences of innovation.	International Water Management Institute (IWMI), Ethiopia	Start: 01–Mar–10 End: 31–Dec–13	Nile
Nile 5	Coordination and multi-stakeholder platforms (coordination and change project).	International Livestock Research Institute (ILRI), Ethiopia	Start: 01–Mar–10 End: 31–Dec–13	Nile
Volta 1	Targeting and scaling out.	Stockholm Environment Institute (SEI), Sweden	Start: 01–Oct–10 End: 30–Sept–13	Volta
Volta 2	Integrated management of rainwater for crop-livestock agroecosystems.	International Livestock Research Institute (ILRI), Ghana	Start: 01–Oct–10 End: 31–Dec–13	Volta
Volta 3	Integrated management of small reservoirs for multiple uses.	Centre de coopération internationale en recherche agronomique pour le développement (CIRAD)/Gestion de l'Eau, Acteurs, Usages (G–EAU), France	Start: 01–Nov–10 End: 31–Dec–13	Volta
Volta 4	Sub-basin management and governance of rainwater and small reservoirs.	International Water Management Institute (IWMI), Ghana	Start: 01–Oct–10 End: 31–Dec–13	Volta
Volta 5	Coordination and change.	Volta Basin Authority (VBA), Ghana	Start: 01–Mar–11 End: 31–Dec–13 Extension: 09–Mar–2014	Volta

Project number	Project title	Lead Institution	Dates	Basin(s)
IF1_MK2	Building provincial capacity to understand the water demand implications of socio-economic development plans in Central Vietnam.	Central Institute for Economic Management (CIEM), Vietnam	Start: 01-Sep-11 End: 03-Jan -12 Extension: 30-Jun-12	Mekong
IF2_L5	Mainstreaming gender in the CPWF: Benchmarking and addressing immediate needs.	Food, Agriculture and Natural Resources Policy Analysis Network (FANRPAN), South Africa	Start: 01-Oct-11 End: 31-May-12 Extension: 1st 31-Jul-12, 2nd 30-Sept-13, 3rd 30-Nov-13	Cross-basin
IF3_L2/L4	Development of a web-based decision support system for agricultural water management of small reservoirs and small water infrastructures in the Limpopo Basin.	WaterNet, Zimbabwe	Start: 01-Oct-11 End: 31-Sep-13 Extension: 31-May-13	Limpopo
IF4_MK2	Sharing lessons on hydropower development processes and stakeholder engagement between Cambodia and Lao PDR.	Department of Livestock and Fisheries (DFL), Lao PDR	Start: 01-Oct-11 End: 31-Mar-12	Mekong
IF5_N2	Participatory video: A novel mechanism for sharing community perceptions with decision makers.	International Livestock Research Institute (ILRI), Ethiopia	Start: 01-Oct-11 End: 30-Sep-12	Nile
IF6_N4	The wheels of innovation: Local challenge funds for rainwater management interventions.	International Water Management Institute (IWMI), Ethiopia	Start: 01-Jan-12 End: 31-Dec-12	Nile
IF7_G2	Implementing community-level water management in coastal Bangladesh: A case study in polder 30.	International Rice Research Institute (IRRI), Bangladesh	Start: 01-Nov-11 End: 30-Jun-13	Ganges

Code	Title	Institution	Dates	Basin
IF8_V5	Volta storylines and scenarios: A mouthpiece for interventions to enhance livelihoods.	International Water Management Institute (IWMI), Ghana	Start: 01-Feb-12 End: 30-Aug-12	Volta
SIDA	Movie on Mekong dam development.	Challenge Program on Water and Food (CPWF), Sri Lanka	Start: 31-Oct-2011 End: 1-Dec-2012	Mekong
IFAD Grant	Mainstreaming marketable innovations.	Challenge Program on Water and Food (CPWF), Sri Lanka	Start: 15-May-12 End: 30-Jun-14	Across
RiU1_PN25	Companion modeling.	Centre de coopération internationale en recherche agronomique pour le développement (CIRAD), France	Start: 01-Jan-12 End: 31-Dec-13 Extension: 30-Apr-14	Mekong
RiU2_PN35	Community fisheries.	WorldFish Center (WFC), Malaysia	Start: 01-Jan-12 End: 31-Dec-13	Ganges
RiU3_PN37	Land and termite management.	International Livestock Research Institute (ILRI), Ethiopia	Start: 01-Jan-12 End: 31-Dec-13 Extension: 31-Mar-14	Nile
RiU4_PN65	Groundwater.	International Water Management Institute (IWMI), Ghana	Start: 01-Jan-12 End: 31-Dec-13 Extension: 30-Apr-14	Volta

Index

Note: Locators to plans and tables are in *italics*.

Nile BDC: adaptive management 58; development challenges 89; Ethiopian SLM program 57–8; gender research 61; participatory videos 67; rainwater management game 67; reflection workshops 67
Nile team 206, 207, 214
Norman Borlaug Award for Field Research and Application 109, 180
Nyando Basin (Kenya) 133, 145

ocean acidification 17
Olifants River (South Africa) 135
Olsen, R. 59–60
Ongsakul, R. 130
online discussion groups 64
online surveys 86
organizational incentives 172
outcome-based R4D 9, 10–11
outcome logic models (OLM) 51, 53–4, 158
outcomes *101*, 118, 130, 171–2
outputs 100–1, 118

participation: communication processes 59–60; in decision-making 60, 184, 209; gender 61; importance 134; plant breeding 117
participatory impact pathways analysis (PIPA) 53–4, 71, 84
participatory theory 174
participatory videos 67
partnerships 55–7, 93, 156–60, 174, 204 *see also* institutions; power relations
Partners' Working Group (M-POWER) 146–7
pasture 181–2
payments for ecosystem services (PESs) 133, 139–40, 142, 209
peer assist sessions 94, 200
peer-reviews 128–36
Penning de Vries, F. W. T. 135
people-centered approaches 7
performance indicators 112, 113
peri-urban farming 26–7, 78, 113, 179
permits 108–9, 180
personal contacts 142, 191
Peru 112, 139, 195
pesticides 26

PET (potential evapotranspiration) 19, 21 *see also* evaporation
physical water scarcity 18, 22 *see also* water scarcity
planetary boundaries 17
planning 142, 167
plant breeding 32–3
plant-field-farm system 4
plant transpiration 19
platforms 160–2, 174, 213 *see also* engagement platforms
polders 25, 104–5, 191
policies: changes 102, 205; dialogues 185; engagement platforms 161; influencing 57; interrelationships 111–12; reforms 118
policy analysis matrix (PAM) 194
policymakers 142, 174
political contexts 148
pollution 17, 26, 130 *see also* water quality; water-treatments
Poolman, M. 135
population growth 16, 18
poverty: Andes 193; and development 7; economic classifications 33–6; and power 163–5; profiles 194; reduction 130, 145; traps 23, 209; unreliable rainfall 23; and water 8, 33–4, 88, 202 *see also* rural poverty; water poverty
power 58, 161, 163–5, 208
power relations 60, 167 *see also* partnerships
precipitation *see* rainfall
private investors 207
problem definition 102–6
production 210
productive water systems 133, 140 *see also* water productivity (WP)
program evaluation 51, 68–9
program information 64
projects: governance 147; influencing decisions 137–41; leaders 55–6, 88–9; monitoring categories *114*; outcomes 84, 171; partners 141; research outputs 102 *see also* basin teams
Project Andes 1–4 193–5
Project Andes 2 25–6, 29
Project Andes 3 25–6
Project Ganges 1-5 191–2